Czech Girls ET Contact

With Humanoid EBE Olie

By

ILona Podhrázská and Ivana Podhrázská

Forward by Bret C. Sheppard

Copyright © 2017 ILona Podhrázská and Ivana Podhrázská

TABLE OF CONTENTS

Forward..1
Introduction..5
3 February 1993..3
19 March 1993...25
27 march 1993...27
25 February 1998..31
15 November 1998..35
15 March 2003...37
12 April 2014...41
7 September 2014..45
25 December 2014..47
26 December 2014..51
6 January 2015..57
27 January 2015...61
24 April 2015...65
2 September 2015..69
22 November 2015..73
1 December 2015...75
3 January 2016..79
21 February. 2016...81
3 April 2016..85
19 April 2016...89
29 May 2016...91
10 August 2016..93
30 August 2016..97
18 September 2016..,101
22 October 2016..107
14 January   2017..113

Conclusion 120

Forward

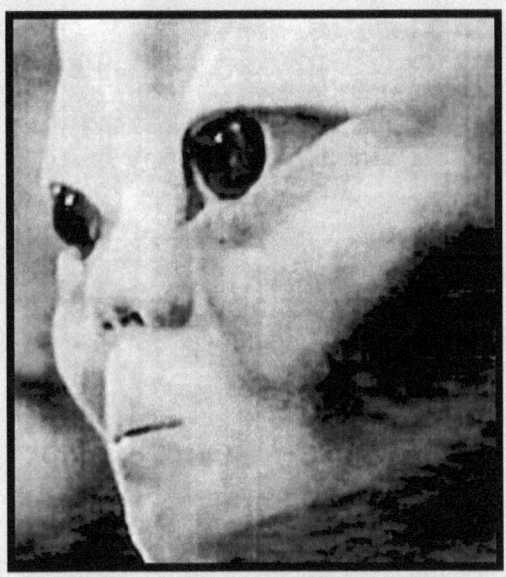

(EBE) Energical Biological Entity

In 1993, Ilona and Ivana made contact with an Extraterrestrial from ELieLjíl. This is her personal human experience. Ilona, along with her sister Ivana, live in Telč, part of the Czech Republic. The area is fascinating because it's near the Danube River, with a rich history that goes all the way back to the Vinca culture. This is the story of ten thousand years of contact. The Vinca culture, the oldest civilization of eastern Europe, included ancient Illyria. The Vinca art and writing directly shows that something happened back then, a possible ET incursion.

Maria Orsic

Maria Orsic, was born in Vienna. Her father was Croatian (Illyrian) and her mother was a German from Vienna. She was a very talented psychic, conscripted to work for NAZI Germany, who was in contact with ET beings as well as other Nordic beings from Aldebaran. Curiously enough, she was also in contact with an ancient race of Earth beings who lived beneath the Earth, or the underworld, as it was called by the Pelagians from ancient Dacia. The Dacian's called this Earth-based race "Pelasg" or "first man."

In North America, there are many Native American stories of these early inner Earth-dwellers sometimes called the "Ant People" who helped them when there was a major global cataclysm.

As we can see, there is a rich history of contact with beings who were already here before man arrived on the planet. Whether humans were created from genetic material, came here from space, or in came spirit form is any ones guess. The messages from diverse sources, from ET contacts with humans, contain information to explain the material that makes up our composition, and how we might have been created or manipulated. I, for one, am a believer in human experience for understanding, as there is nothing lost or gained by simply believing for the sake of investigation, until proven false. The automatic writing in both cases from the Vril Girls and the Czech Girls is similar. This is shown by the fact that EBE chose to communicate in their ancient languages.

Maria Orsic's writing was analyzed and first thought to be Sumerian, but that is not accurate. The writing was from the Slavic/ancient Illyrian languages. This would have Ilona and Ivana's automatic writing sessions make sense, knowing those writing characters originated from the ancient Vinca culture, from the culture that lived along the Danube river, where all of these girls were born. The people with these ancient roots, from that particular area in the world, seem to be favored by ET for many reasons, one being genetic memory. No one person is smart enough to handle all knowledge independently. We can put things together that require the expertise of those concentrating on a particular field, but many times the messengers can only communicate these ideas and pass them on to someone more knowledgeable, perhaps just for the future. The Vril Girls and the Czech girls are mediators of this knowledge, that is their expertise.

It demands that we do not become lazy in interpreting the knowledge that comes forth through them, or any human used for a portal of knowing. Phrases are being fired out through their automatic writing that make sense to others, to scientists, geologists, and artists that can see the other side of the status quo. We need more attention on this type of phenomenon to make our world a better place for all people. At the moment, these things are seen as eccentric parts of our society, to be ignored.

Extraterrestrials see humanity from a different perspective and ET races are trying to help we humankind, stuck in the material world, and they see our separation from source (cosmic energy). This is happening through ignorance of the natural information from the collective, information from people who live and breathe on the other side. We should work together, with integrated knowledge, to solve the problems we have made for ourselves because as a result of ignoring Nature within ourselves. These messages are real and true to what I know.

- Bret C. Sheppard -
Curator of this project.

Introduction

UFO Congress

Czech Olomouc – 1996 : Speaker : ILona Podhrázská :

My name is Ilona Podhrázská and my sister Ivana is a medium to communicate since 1993 with Alien, ETs EBE . " ENERGICAL - BIOLOGICAL - ENTITY " with " psycho mental methods spiritualistic techniques . It started invoking spirits, and then contacted At least you and say that they are aliens from the planet ELieLJi . I want to be with you share. Like me read concisely information that we receive from them .Classic spiritism table. We paper thereon circle with letters with letters and sister touching, touch small glass , which runs, go after the paper beginning .write ,go letter. Ear lier EBE write wrong Czech. lately improved. Sometimes I write some sort of CODE. So I will read, what they write what they tell . Sometimes They write sort of Code ! It will be very concisely . It all started 3.february 1993. Messages - messages from alien EBE - OLIE. I 'm not a ghost, I'm an alien from the planet ELieLji .

EBE Olie is our great friend from outer space! From the beginning he wrote a lot of errors in the Czech language. Often used for foreign words .In our over time EBE more improved. There are many extraterrestrial entities in the universe. There are many people who have various contacts with aliens, and there are many people who are afraid to talk about it in public. I am with my sister, I decided after several long years to go to the public, to provide this information to light Is there a lot of people who do not believe us, they laugh at us and defame us. Everything is controlled by a higher power in the universe .Neexistují no coincidences. Everything is in the program ... I know that I still have a lot of work. EBE once wrote: "Contact does not occur quickly, but slowly and deliberately. Everything has its time with which revolves the whole universe and the mystery.

I have such an experience: around 1994. It was interesting that when I woke up ,, I went into the kitchen and my mother noticed that I have in the back of the nightgown burned differently sized holes. There were many holes .. Then, about two hours after I stood Ivan, which has a room next to my room. Ivan woke up and had too much on her nightgown different sizes, burned holes .Vzpomínám think that night I was sparalyzovaná, which happened to me often enough. Now when I heard I seemed to paralyze morzeovou alphabet: Is it weird that we both have had the same day burned holes in her nightgown. I once had a "vivid dream" that I was lying in a white room on a table in the shape of a hexagon and around me was about 5 humanoids, and I was lying on my back and I could not move. The next day I came back to my throat burned like two punctures red side by side for about 4 cm. apart. Two punctures on his neck, I had clearly seen for about 10 years. Over time, it disappeared. Direct contact, we have experienced many times ... once in the year. 1994 we communicated with EBE Olie about 30 people, including one priest from Znojmo in Oblekovice. There was one lady MD. Alla from Ukraine, which says it is so. "Trans Cosmology" MD .. Alla Kulikova from Ukraine. Gathered around 30 people. Everyone was curious what will be Ebe write. EBE then wrote Ilona, Do you believe - believe - neutral). Ebe wrote that in 23 hours, we have to go all out on the court and that "they" will show us so. Manifestable years. The former were thick clouds. Alla Kulikova said that we can help. Alla began to focus heavily, got into a powerful trance, she began to tremble. One hand held the fourth chakra and the other hand was pointed up to the sky! It was "Atmokineze." - The movement of clouds with psycho - energo - sugestologie. We were all in shock! Suddenly, the clouds started to form a gap that more and more expanded and spread up to heaven in a large cleared. For a while we all waited and then it occurred. They showed in full beauty! Their extraterrestrial object flew zig-zag back and forth and forth and back and forth again, and it showed 7 X back and forth. Their object glowed green, and then a little red remember that the people there wept and said: "They really exist !!! They were very surprised by that. And I with my sister Ivana We were also very surprised from the fact that she could Alla limber with clouds. This is one of the experiences that we have experienced with humanoids ... of course we have more experience. and if my time allows, i would like to eventually wrote a sequel and other information about what we humanoid Ebe - Olie telling and still conveys. In this book are only just some of the dates with the information that I have just herself translated into English. I'm just a self-taught English only a few months. thank you very much my great friend from Facebook Name Bret Colin Sheppard! Bret is great researcher, researcher, writer from New Mexico. Bret information from Ebe very interested and so I Bret offered that I would very much like to help create a book. I agreed. Gradually, I sent him a message from Ebe and he's put into the book, which is since February 2017, published in America! Our first book in English! Now, many people wrote to me, whether I create a book too in the Czech language. And so, together with Bret We created this book. He offered me a translator and Stuart Spencer from Brazil that would love this book translated into Portuguese. We are very grateful for the creation of this book!

- Ilona and her sister Ivana.

Automatic writing of Maria Orsic

Ilona and Ivana's Technique for
Automatic Writing and Spiritism

The following is the original transcript from the automatic writing communication with EBE, Ivana, and Ilona. Ilona states that "Ebe from the beginning, wrote in the infinitive and often used our foreign words and various codes." For reasons of authenticity the following text is translated to English in it's original state by the sisters and have not been edited into a more conventional form of an English dialect.

Definition: Automatic Writing

"writing said to be produced by a spiritual, occult, or subconscious agency rather than by the conscious intention of the writer."

I'm not a ghost

3 February 1993

I'm not a ghost, I'm an alien from the planet ELieLji . "In America we say EBE" . " ENERGICAL BIOLOGICAL ENTITIES". My name is OLiE from planet ELieLji " We come in peace, and we want to protect your planet Earth from nuclear explosions to prevent the destruction of the Universe. We are glad that you believe in our existence. Ivana is endowed with special powers . Ivana is the mediator between Earth, and the planet Elielji . The transmissions are a kind of navigator through a computer transmitter, to communicate with people at a distance. We are humanoid like you, and we are out of power. Our existence is a way to level the existence of the other spatial coordinate . They feed with energy. We enjoy music, and we are in the heart of the chord melodies. We have 7 colors like yours. Eyes color we have caramel , arms and legs yellowish . We have 4 fingers white. We merged the hydrogen molecule . The molecule is a concept that is found in most Universe and the organs. They feed on just energy. We do not know what it is hungry. We do not know what it is when it is wrong.

WE FEEL STILL NEUTRAL . We appreciate on your kilograms.20 pounds and measure a 1meter 20 = 1,20 meter, We have liquid that is white and consists of a hydrogen molecule . Body Temperature and September temperature 45 great heat. Our body gets hot. We hear telepathic wording , it is into the computer technology . We move from the ground a few centimeters. Our body construction holds only matter, energy mass. Plant has a structure like root that holds it. Holds us root mass. It is the root, who can not sow , but which originates from a cell. WE mutations on the level of technology. Hybrid creation is at the level of biotechnology. It would take more biological mutation associated with us ! It'll STRAIGHT - humanoid existence. The pyramids are the base of our past . The pyramids there are holes invisible to you. That we secure . Fly to clean technology ENERGICAL SYSTEM . ... We are of the dimension 12 and 39 light years distant. Distance does not play role. Move " Spatial temporal jumps. " We have coded your gray out here on Earth. Everything is recorded in the procedure. You say "UFO " . This is called in another way . 4:06I'll write it on a human level , " CONSTRUCTICAL - TECHNICAL - NITRICAL " . We we fly with technology and with the design , to "nitro "INSIDE - INTERIOR" for flying. On the Moon, there are runways For among the - years ago. Robots are navigation extraterrestrial creatures. Robots are also transmitted together with " Konstruktikal Technikal Nitrikal Constructical - Technical-Nitrical " on Earth . Robots also examined the surface of the Earth and your we were taking it with us on the computer . The robots are designed to explore the planet's surface and we examine people. It divides the Biological and technical abstraction " AVATUTA " . We investigate people. Investigate people. We take blood samples for research. We need to have the device in the closed state. We investigate and abdominal openings. We do not want to hurt anyone . REGENERATOR - unit is to examine the animals.   yours NUCLEAR PARTICLE penetrated into our bag - SPHERE. Also whit me sometimes loses telepathic waves . Among all - star of the runway on which our alien ship . Our planet can not see it under you guys very much . But the track is equipped with a quick transfer of our space rockets to you here on planet Earth. We have other galaxies. Ivana has its headquarters in the brain recorded point here for us to base. It 's not that we He appeared to more people and also fear . Everything needs time and logic . It what the Ivano and Ilona feelings so -called vibrational energy . It is a destination for early contact. Contact does not occur abruptly, but consciously . CONFEDERATION - the law is satisfied to different situations. PES - our navigation . Marahata Charara - our squad UFO. GADA is one of the flying shuttle. ŠADAMAKA - Shadamaka - EARTH.

Ivana's spirit board

When is less genetic experiments , the better. it will be Development through electro-magnetic wave measurements from a distance and up close. ELECTRO- MAGNETIC WAVES WAVES has no effect on poisons and what they could do harm as genetics . Is it time MERA ENERGY Electric - MAGNETIC MEASUREMENTS 10 or more units of frequency calculated in the magnetic scale. It has a range of 20 frequency units and the way to a total of 30 units of frequency and it counts it as the end of the experiment. NUCLEAR RADIATION fuels at you also leads to damage. Does this method of radiation radioactivity and the total terrestrial temperature. The temperature varies with you that you have broken earth layer of the black hole . people are of that groggy . PLANTS DIE OVER THE GREENHOUSE EFFECT earlier. Ultraviolet radiation is also bad . Telepathic RADIATION - It's a match hiding information from the subconscious And in any case, the blockade of space-based communication . WE HAVE SPACE LABORATORY . to base many units HEALTH front. Into this SPACE LABORATORY I ARE SOME RESEARCH Earthlings taken away to secure for health. inducers - a thing which can be used to establish a connection with the Earthlings, as a combination of reflex. When earthly man is under our building, so it means that it is programmed , to get to the object's angle increased . Get him in Nitro " inside - interior" simply using lazer, which is strong, exudes power and magnetism of the man is drawn into the interior of the building. Color elevator is yellow- white-blue In our system include the planet : Mursa, MiFia , GUMPOLA, POLIZARKA and RADAN .

The planets orbit around the sun, but from a different GALACTIC CENTRAL PLANET . They have a fiery color , just RADAN is white. Ilona, Ivano ! Be careful in Then what are you doing ! It is a large telepathic communication hub overall perception MY humanoids we ether   NEAR KONSTRUKTIKAL - TECHNIKAL - NITRIKAL ! We have some protective devices that attacks against infection . We have the OPTION BASE TO START A DRIVE Seconded energy streams on Earth. From such a distance , WE ARE ABLE TO RECEIVE DIFFERENT audio transmissions from human BIOLOGICAL COMPUTER   . We have " Sounds - voice adapter " RADIO adapter So and that and that the VOICES are transferring into our MACHINE " KUBOZA " . EVERYTHING IS SCHEDULED AS PROGRAM on the computer. Not to forget anything ! Man is the most interesting part sample. The samples are piped to instruments that measure their valuable RATE. PATTERN IS USED AS A PARTICLE TO EXPLORE . SAMPLE each person is different. Such information communications are part of a trial and not to information so that someone misused. This is information dignity. his is not a fairy tale, but the reality .. CIRCLES in the boxes are wonderful people to do more and more aware that these marks were left behind other beings from other planets. Is that a flying object landing BEAUTIFUL . From the body can be established from the flight CONNECTED PEOPLE that are necessary PARALYZING

or can be stationary object , such as , for example, auto ... All this can affect the cosmic body . FIGURES IN THE FIELDS message. THE FIGURES SHOW MUST be more celestial bodies to develop a more geometric PARTS . BODY TO CREATE A Sorts Beautiful, ALWAYS BALANCED shape. After the flight speed creates energy that is there gear. , That there was to see. The body of the flight , such as combustible material that will not burn, but it is hot the flight speed. When people see each other more bodies behind , so that's it. These multiple bodies that even the sky is a sort of a set of patterns. Often the planet " Zeta " and the planet NATORIA . It is enough proof that visual in all kinds of places in any part of the planet Earth . Your planet Earth was habitable in ancient times different civilizations which left here on earth also enough evidence of tangible and spiritual . For example Monument and pyramids, moai statues are also built ancient other advanced civilizations. TERMINATION OF PLANET runs in constant change -TRANSFORMATION ATOM With PROTONEUTRON PARTICLE- . Is chosen, that when the joints in the stratosphere ATOM With NEUTRON THEREFORE , it is without a doubt the Planet POSSIBLE . NUCLEAR ENERGY IS THE COSMOS And the existence of cells in the cosmo- SPACE. The universe was able to bind cells to the good life. neutron PARTICLE COMBUSTIBLE PRODUCT ARE IN THE UNIVERSE . ARE TO SPACE particles. BUT I ON EARTH Is neutrons and protons , BUT NOT THE WHOLE PROTONEUTRONU monster . Metals are the worst Total unit - whole for the maintenance of clean energy. KONKURÁTOR -" CONCURATOR"- is with you, as a probe . In our t's just a thermometer and pressure. WE ARE ABLE TO RECORD RECORDS CONCURATORO temperature and pressure on the country. Just off our ship

long distances . A PROBE TO YOUR FLY just around the earth's axis and follows it's own larger space Z and the Middle WAVES monitors temperature. And we can do this enclosure and watches from a distance. The whole universe is as the single largest computer program in human life.s.
PARGANÁDA - a development program in human life.
MUSILIKACE - is to move ideas into the past tense .
BUDISIKACE - is to move thoughts on the other side of the ordinary . CHOREOMAGIE - is a system of ideas in human knowledge . RELACTOR - the arrangement of ideas that relate to parts of the spherically arranged That is, even your thoughts are set as a signal to the Universe. between the brain and the spherical mass atmosphere up to a certain wavelength mass of air and atmosphere are A PARTICULAR INSTANCE DEGREE REACH - bland whole .
ASTROSPHERE UNIT , which normally One can not pick up material, but as practicable . Just from measurements from the probe. I love the Earth , and you, but I do not love war , destroying your planet Earth, and soul War is a poison that has entered into BETWEEN - Galactic expensive. Man is a man without emotion. often make Examining the blood or viscera, but that do not harm anybody . We evaluate the whole process of your body. We accompany it certain DEFINITE MENTAL PSYCHO magnet that pulsate invisible process. It's for you, as sent satellites that will also appear pressure in the atmosphere. So he MAGNET No one in order not introduce you to radioactivity . No - why ? We intend to HOLD terror and degenerate living cells, the planet and humanity. BIOLOGICAL institutions include pollutants in cells

LIVE VIABLE. This biological GENERATION IN YOUR COUNTRY WILL NOT live for many years. Cells are impaired ! EBE is like , I'm proud of my existence. LAZER CONTROLS the ROLATION of the Cambrian in Hydrosphere BY RAY LAZER runs in the hydrosphere. Cambrian TO THE CHEMICAL COMPOUNDS . I Olie I have become critical dimension for a given potential. On your planet for a given potential. On your planet radioactive pollution. In your vegetables absorbed carbon dioxide and nitrates. It 's not so good for the human body , but it's still more healthier than milk products and especially meat. Proteins can be replaced otherwise than meat - cereals EARTH disaster is a big mess ! I have no ruler control for order, for the planet. BRAIN IS THE MOST IMPORTANT MUTAMENTAL BIOLOGICAL PARTICLES creation. Instead , it is interesting to study Earth's knowledge so it is Newzealand , Mexico and Cuba . Of course the patella. Yes , there are examining what is identical with the soil and plant We compare BIO - PRODUCTION CREATIVITY. Make.In Costa Rica underground pyramid. Soon you will be struck between - the planetary system in the contradiction between earthly and OTHER SCIENCE . It will be a proper parliament telepathic and direct sharing , I think the community. One will be , you solve with your science. IT'S UP TO YOU YOUR FATE I PUBLICITY moral conduct. WILL HAVE TO TAKE PLACE AT - planetary base , I have there other civilizations .. IS THERE focus telemetry and graphical sound system . PART V protons neutrons CONSTITUTES THE STABILITY OF ATOM in space. These are the positive and negative particle sources, which holds a large number of atoms in the universe. STRUCTURE OF THE biological organisms is essentially more complex than the composition of the matter in the universe. In comparison to the human organism is with us again on a different principle appropriate to life. Our liquid, if put in comparison with your organs is so much more than human organism. Replaces the body more. BIO - ENERGY FIKTRACE! Fictrace! depending on the core biological part of the Universe. It is the cycle REMAINING SPACE sporadically ELEMENT IN THE UNIVERSE, pulsates between them, such as magnetic Convention. It is connected to a signal from the base SUBSTANCE CONTAINED IN THIS particles are energy based on the way COMPONENTS REMOVAL inappropriate poisonous gases from the ionosphere particles in AZ and from NUCLEAR PARTICLES ENERGY that structure BURNING THE BIOLOGICAL PARTICLES would result in a bang at that particles and it would break part of the atmosphere and the biosphere. Biological particles from bang in the universe RESULT VIOLATION OF BIO - atmosphere. BIOLOGICAL ATMOSPHERE IS based on biological particles in the atmosphere. MAGNETIC CONVENTION - serves as a substance contained between biological and energetic particles among which pulsates. When paralyzation sometimes hear strange humming. PROCESS METRIC wool proceeds with each of earthlings. IT'S WOOL, FLOW OF BIOLOKÁLIT in a living organism, to purging, cleansing cells. When the organism dies, is dead METRIC WAVES so does not take place, but in the last moments of ongoing leakage of cells. Lead cells into one mass.   MALALUJA - Entria- Sutus - Lapacha - LUNA - orbita - This is a situation at an advanced stage of development of biological components in the inter - stellar SPACE. This is our speech. We're talking strong, but you incomprehensible language. Terenty - LUTAS - MITORA interference basis for use in terrestrial beings. Ilona, when you hear the beep when paralyzation, like tuning a radio so it is a signaling attachment point on your perception of our approximation. paralyzed by the contact of two energies in a single energy that flows armor to immobility. to radiation energy. I will have the protection that is my law Take care of yourself ! Time for you on Earth passes quickly and that we are approaching our meetings. You can see that we can not give up, stop communicating with us. Stop being with us and feel our presence. Vibration earth's axis are strong. We see it on the COMPUTER. Your planet has a change of climate. Solar radiation is harmful to every poor individual who does not have the right blood circulation. RADIOACTIVITY AIR is rising, capturing POINT OF RADIOACTIVE SMOKY THE COMPUTER. It is bounded by a kind of shutter covers. Style in the cosmic understanding of Earthlings is unformed to higher susceptibility and higher knowledge about it. The foundation of positive energy transmission from human to human is essential. The idea is the source of telepathy. SIGN IN FELDS WILL EXPLORE MORE DISCOVER. AS THE ARE DESIGNED CODE THE PRINCIPLE OF TIME! And CODE has its your time ! Everything has its plan to be the rotation of the universe and mysterious. The Force is all around you, which acts as a magnetic pulse. Our commander is called TALAS AZULER . WE HAVE A LOT OF WORK FOR THEIR PLANS . Offered PLANS BASED ON BIOLOGICAL SAMPLES FOR UNDERSTANDING BIOLOGICAL SCIENCES on your planet Earth . Thank you for your attention.

Ilona writing EBE Olie's information, received from Ivana

God Created all the Worlds

19 March 1993

ALELUJA OO ! God created all the worlds and us, and we have assisted in creating you. We started with using robots on Earth and then biological artificial mutation, and then the development of the organs, then we were unable to complete a higher intelligence, as are we . Because of what we were doing there, we carried the knowledge to the planet but soon realized Eliel and we could not complete the creation process. Animals and insects got here through experiments similarly like you. We on Earth had two days as there were giants, and we had just a short time too create. We have a fast technique. In Costa Rica there is a base from space. Your Country arose from the disintegration of a moon, and From that moon began to fall a thick red substance from which it will then form compounds of hydrogen, nitrogen and oxygen, and then the mass of the atmosphere. Note: The Red rain incident of India.cont. Then after, we discover and explore, and look after the intention set into motion , We only had two days to create. Then we left, because we were low on fuel. We had to leave the earth at that time or we would be stuck there, so we left the creation undone with a higher intelligence.
Q. Ebe what I have in my ears?
A. Earring .
Q. What does my mom have on her thumb. ?
A. Spool.
Q. What is the "eyes"?

A. Peepers.

Q. What Ivana keeps in his hand?

A. Sneezing thing .

Q. What are you most afraid of?

A. Nuclear weapons . ILona, please do not ask questions about this.

Q. Where is your planet?

A. Zodiac, Scorpio direction.

Q. Have you ever been to Mars?

A. Yes, there we were exploring scientific findings, and stone archaeology. Mars is similar to Earth like your Adam and Eve. In Costa Rica, the pyramid radiates a strong energy. Ivana has the strength of a snake. Ivana is sleepy.

27 March 1993

Q. Ebe, landed UFO somewhere now?
A. Yes, in Ecuador. and Cuba on February 7 1993.
Q. Do you have hair?
A. We have light short fluff, the color of snow.
Q. How much of our species is extraterrestrials?
A. Enough.
Q What is the composition of your body?
A. Energy, mass , like a plant. Ivana will be carried away. We will give her lungs and examine ..O Ebe Ivana career. I do like the needle. You have our abilities. Sir, you're like a zebra, black and white. "Okay", meaning we are close to your tightly connected energy.
Q. Who is this David Copperfield?
A. A terrestrial Mars man . There was life on Mars. We Ivana have examined the lungs of the universe.
Q. And why Ivana?
A. Your mother had no pain. Childbirth is easy for you. For the past 7 generations. Your ancestor also had a great ability. Some People are glad that this sounds crazy as it is crazy advertising for cover ups. You're very curious to us. It is true that we created Adam for the swarm (Hive)
Note; Cybel, Bees,male clones.
"Peafowl, RACHEL" . Guided brain walking . . "MAHA PATEL ZREMALA". Ebe piece is the poppy head . "Greeks - pit - snake" . Ebe knows it like the Latin. "Small revue rustle Rachel."

Note: Rachel is a biblical character in history, and the mother of the people of Dan, Danube, Dana. The reference also is an insinuation that religion was coded in our brains.

Q. How do you reproduce?
A. Needles. Seed.
Q. Where will you land?
A. Sweden.
Q. What is the brain?
A. It is a Biological computer.

Ebe gave us proof! In 1994, we communication our classical technique spritistism with Ebe. There we were invited to the Czech town of
Znojmo from the lady of Ukraine healer Mudr. Alla Kulikova, which can move the clouds, known as Albert Ignatenko. "Atmokineze" She met us there about 30 people. There was also a pastor. Evening communicate with Ebe and Ebe and wrote: At 22.30 hours Go out into the yard and watch the sky, that you did not think that we have some ghost and extraterrestrial posing as ... Go out into the yard, and we demonstrate demonstrative years. Damage, But the clouds. Mrs. Kulikova did atmokinezis , the clouds disappeared and we are 30 people watching the sky at 22.30 hours. And suddenly it was a big surprise! Enough to clearly see the object with three lights, two green and one red. The lights circle rotation. 7 x object flew back and forth. People applauded and said: They really exist! I saw 2 lady wept with joy. The pastor was also very surprised. This was proof to us

that Ebe humanoid it's not a ghost, Not he makes us laugh. These surprise we of course had more. We recently communicated before one Air Force pilot, who lived in the Australian outback, now lives in Prague. Ebe suddenly began to write french. I asked the pilot what it is worth mentioning what is this language? A pilot replied that he thought something in French and French - Ebe replied. This is also very interesting. Ivana and I do not know French, and Ebe wrote French. In year 1992 I went with my sister home in the evening and I open the door and us at the door to the backyard was a small humanoid, I startled him, startled. It really took off and I quickly closed the door. Ivana and I saw it. Then we hear strange noises. We hear from above, from heaven like a male voice called my name 3 times Ilona! Ilona! Ilona! We had a lot of fear and quickly we flee for help to the police. Police drove us away in front of our house and our mom is in the window and talk: What is going on here? What happened? Police search and find just dropping, broken vases for flowers. Then our mother remembered that about an hour before it was Mom in the kitchen and heard strange noises. The first thought that I and Ivana are in the yard and that we want Mom to startle, frighten. Mom got scared and went into the living room. Since the year 1993, we still starting communicate with humanoids. Yet sometime in 1989, we have myself in the winter evening I went to go for my friend to visit Marcela. It has been much snow, snowing. And in front of me I went too high about 3 meters tall as a man and had too little head .. He certainly went. Steps like a robot. But I first thought that it was someone like fun and go, as after stilts. He was very tall and a small head with a hood, .After the pavement in the street walked a gentleman and he said: What is it? This probably is not a person. I too was worried, but I still thought that someone was joking. Then, in about 22 hours I come home and my mom says Ilona, 'You're not afraid? A while ago spoke on television that Russian UFO and there were tall aliens. Well, I speak: That evening I was too much of one extraterrestrials vysokýho about 3 meters I met him on the sidewalk and in the street that I thought that someone was joking. It was too much snow and the creature went too sure, she went straight up, like a robot. Steps like robot.To for us mystery. Often myself and Ivan were paralyzed. One paralyze when I heard a beep, like Morse code, and heard a radio tuning. Then I got up in the morning and my nightgown was full of holes. Burn holes. And Ivana was sleeping in her room in the morning when she got up too also same had burn holes on the nightgown. Holes burned in the back, everywhere small burn holes Strange that we had both same. We have many mysterious adventures.

Area 51

From Ebe -   OLie : Area 51 - info:
25. February 1998. Communication with Humanoid EBE-Olie from 12 dimensions. Through spiritism   methods from Czech telepathic contactee Ivana Podhrázská and Ilona Podhrázská 25th February 1998 - ALeluja oo ! ! "There's no need to rush into everything. The fact that we did not communicate for a long time nothing is lost. There are many people who are afraid of us and think about us that   we are spreading evil and other things on your planet. Can the people   are afraid For ILona a Ivana that they transfer the bacteria from us. Those who will believe us humanoids communicate. See, people   to you avoid and gradually. But do not worry! We know that   for us in   laboratory technology in laboratory   bunker on us doing   against us weapons for destruction. They want to destroy us. Do not believe it, how to write that the Area 51 aliens keep maintain secret research. This is not so far from the truth. At Area 51, they are just trying to adapt to our technology. They want to create something that could ruin us in case of emergency. . They , only   so think . They themselves the scientists register above their base Area 51 and they watch on theirs computers,   how our Ships   and   other ships work and how it develops. We are afraid of other   beings are afraid. But we know that still   there do not nothing completed. They fail to do it. They do not do it . Does not work. Not function . It is true that there are also materials, but not from your world, but from another world. According to the various

iron alloys are they try on effort to create anything resembling , similar ours machines of the same material. With that technology may threaten humanity. They can attack. It may have more danger. I can not tell you everything. You no must nowhere say it too, because you do not know anybody. They can work with them and the people who know you . They about you may also know, because they have different records all contactees people from around the world. They have it on the computers all the data. II will no say thee Ilona, Ivana, whether they people of Area 51 they know about you Ivana , Ilona know or not. not sure know or not know, but it's possible. Therefore rather keep secret such news today. It's no good . Because what is concealed is always dangerous. Area 51 is a secret. And why ? Because it has its purpose in case of any danger from our side. So that they could intervene. Or would that they could be used in wars conflicts . In Area 51 there forms a weapon based on technologies outside of your technology and such bases are many. If you were closer to the base, you would not escape, because they would you have hold there and they would like any information. No escape . And also the reason that you would be if you're about to talk somewhere public what you saw there.So this right, we explored these days. In Area 51 there exists great confusion. Every person there has a lot of papers, documents and elements. All possible sorts of materials. It is also a chemical origin and vegetable origin. What people have for the world, when they collapse under their feet? It says to you., we already know the meaning.Your Earth , there is just

Reigns only the financial aspect. Ivana, the money, what you miss with that we will not help. It we no have in the plan . For us it is illogical that money ruled mankind. I have already weak information. Only I tell you, Ivana and Ilona , better stay keep in a closed circle! Yes, already i will end , declining, diminution energy. Next time we communicate! ALeluja oo! "

Genetic manipulation

Ivana Podhrázská and Ilona Podhrázská from Czech republic :   Excerpt Message EBE - OLie from Humanoid Genetic manipulation

15. November 1998 : " It is true, that we take   removing , for men and women samples. We examine whether it makes sense. It is an ancient system of the origin of life. But Hetero - humanoid race is and will be normal. They will not develop people, but half people   and half of us. Hybrid offspring to save the planet Earth. Therefore, not end of the world. This is all planned and so it goes. The end of the world spread people and   seers   who fit the Jehovah's Witnesses. In the future, both sides genitals longer needed between two people. They will be connected to the device live biological particles. with the cell. It will form in the vacuum of space. Must surrounding zone cleanup. There shall not be contagion. Otherwise, the fetus never created. We must examine the people, whether they like it or not. Our second group of humanoids from the same planet is still surgery on humans. Caution! Most knows in America. Tissue human   are ready to stab into the bodies of our colleagues. Nobody's hurt. Then even the sexual organs is injected. And then left the vacuum of space, and biological device ..... born a mutant. It is not a child in worldly woman, but in our beings. Yes, it's too much. They live on the planet and are preparing for another technology with you on Earth. And that just want to prevent the secret service. They want the same thing. Your secret services they want from people making mutations and want our abilities. . I can not tell about it. It is in great danger. It is, as you do on Earth, once the Cold War was . Everything we have on record and how many people were killed mercilessly, which is a shame. Your planet is experiencing a hell for many years. I'm going to go, I'm busy. Gotta keep the program over the Eletrictal factory / Ebe and where are you ?   / -   Over eleelectrical factory three kilometers. I do not know if I can write, Ivana wait and Ilona. I'll check ... Yes Russia. I Gotta keep black box in a program that shows the status of the affected environment. Surely you do not understand the concept. . It is our origin. For you nothing like this. . Just a plane black box, but it's different. . It is in the program. Records ozone and black draws him into her. Then we have the computer opens in the program and we see through a large image of a black, gray smoke. We measure the energy of uranium and We spread on the elements that exist in your country. We scares what your smoke contains everything. I'll finish. ALeluja oo!

Humanoid Suits

15 March 2003

Communication with Humanoid Ebe - OLie manner spiritism method telepathy from Czech republic . ILona Podhrázská and Ivana Podhrázská ... / Question : Ebe, you have suits ? / - Yes, we have suits that protect our contact with the atmosphere. Coveralls isolate hazardous gases. If we were not protected by the gases we would attack our body that we have, although simple, but it would mean tissue damage. / Question : Who manufactures overalls? / - We do not have fibers . Coverall is made of our resources. Of the various alloy elements. Not having to production . We do not have , what's with your on Earth . We do not have factories, store, shops . We are creating this through the various alloys , which create shiny, durable isolated whole .

Something like your mercury. Create slowly coverall with using   computer. We have deployed an identification code for each individual. which then manufacture these suits helping to form and shape according by identification. In most the body we have the same, but overalls not the same. There is only a difference in a color shade which create alloy. It is the color of copper. I'm dressing up hereby computer . Computer do   shaped coverall, And according to   code will be produced . And we just using computer all   alloys we do sleek . Until created formation suit, coverall. At us on the planet we are isolated. Protection coveralls we have , just only when researching and investigating samples of your planet, especially samples of chemical and biological origin Question, and you have something on your head ? / - We do not have a helmet. Our head is itself protected and isolated. The head is our most perfect   part of our existence . Our head is like a navigator connected to the computer. But we're not robots. We live like you. Coverall protects us only a body without a head.   Q /-   How do you respiration ? / -   We do not have organs , like you people . We have white fluid , white blood. it's dense. Q / -   Seeing your eyes ? / - We see the spatially whole head. This is a whole unit . We no hear, as your people ears. We do not have ears. We hear telepathically. All functions perceptions we have other . Now at the moment we not have, none task . But we are waiting for our mission commander. State of peace , so far ...but we must monitor what is happening anyway.

Od   12.4. 2014 –English –EBE : 12. April – 2014 UFO - Conference. Discussion with Humanoid Ebe   from 12 dimension - together with Czech exopolitics and filming communication . Camera - chief Czech exopolitics - Karel Rasin.   Subtitles ... Spiritism method .

Chaos

12 April 2014

ALeluja OO! I welcome you together. It's full of energy, Ivana. EBE is another area that you can not see. It's the law space. It's a crisis. Politics have great chaos. He knows how to manipulate the earthlings. It's chaos, as Area 51. Such chaos is for you and in politics in America. Scientists and authorities in different countries are in the hands of confirming the fact of extraterrestrial entities. Everyone knows enough about them, but they must hide. This is the basis, it is wrong. It should go to the light and do not hide it. The atmosphere is at an angle, I guess, " FoLio " is little oxygen for you on Earth.
Everything is currently shooting in our winding system. I have been watching the political atmosphere. Our ship is in the state on a different basis. The ship can not see, but could you see if they joined forces as it would go. Force energy that flows from your hearts, hands and all that mysterious, which is the foundation of your existence. In the hands you have the power. The brain controls the soul of your heart, your hands. The first was to create the basis for the human soul, then there's physical body, which is important. They control the destiny, but due to the soul. HuMan is a creature to a large scale. It is a resource sort POWER ELECTRICITY .
Everything carries its importance in the history of the Earth. Your country has a different name. Country is not here by accident. It is brought to your higher powers of other units of heredity. Cells are hereditary. Everything is identical with the

universe. Love unites us all. BEING infinite universe. GOD IS NOT BEING, HUMAN , BUT IS TO LOVE ALL. It bears the name "cosmic laws" . Either someone chooses the way there or back and logically move away from its fate. Surrendering the chance of survival. Chances of survival are primitive chaos. Our fundamentals are enclosed as much energy for others, as well as other creatures of the planet and the plane was going to happen yet many mysteries. The aircraft is in another sphere, but so far people live. They keep it a secret. It will jevo that he found the wreckage but found. Somehow we must talk. . It is their duty stupid. They do not protrude reptilians who go blind. They are blind, so blind. It is incomprehensible race that is facing the challenge of the government also hush up. They are everywhere. Ivana yes, he knows civilized dimension ZON. Yes they are in your area, but another dimension. Yes, they have blond hair and their decision is cruel, when one of them begins to act. It's cruel, faced with the dignity that people are losing within. It is so. We are neutral, we are worthy and we have a commander who controls us, as your souls of people who died. Anyone who dies and carries a commander and he knows it five minutes before he dies. It may also be the day. The man before death is already informed about his transition. And either will be there or there. Either or. And when you have to deliver what is given to you, what task you have. There is no need to change or dissemble. Then it has consequences in other dimensions transitions. Indeed the soul is before the demise informed and terrorized over or not. We have a lot of work, we get a lot of material, the receiving data from your government and all countries. We must thoroughly examine the ravages of what had gotten out of reason. Must govern higher reason connected with other dimensions. Then open your eyes. People are blind, it's the illusion of this life, but love should be. Goal - contact. We know a lot, but we can not speak too much. It is a cosmic law. IT'S DANGEROUS. We know where the plane is, but it's dangerous. We know it is another dimension and space under the sea, but also in Hydrosphere. It is at on the planet, as a dimension Zon, it's too populated. It is too much. WE THINK MY CALCULATOR. We will know RATE FOR CERTAIN MATERIALS AND CELL BASED ALL WHAT IS THE UNITED. " VINAPA UHAJAKA SALACHA MANLA IECHA" this is a brand that mind. People are always good values together. Career, power, love and chaos . I guess , a substantial'm told. Even the transportation of souls, it ja ta ship what you saw. Ivana and ILona saw it. That was the boat 2 bars luminous and moving to space. Sometimes it shows. Ivana heart charm, is not common. Since time immemorial, her soul discharged on the same frequency. Just body changes in every life. It was an officer. Monalíza also. She had prosperity, luxury. Its agreements are on a different level. It was at the Castle. He stayed briefly in Sweden, France, Italy, Czech, America. He stayed there shortly before she died. She had been sent to a planet that is habitable, but there was a disease that affected the nervous system. Affected nerves in the physical body. It was not a problem to get away, but the technology had long since disappeared. It was suffering the disease. The planet is not there anymore. Intentional explosion. I guess it's all energy by Ivan. Love EBE ending .. You will good and cultivate in. wearing good. This is the foundation of health. The cells will be fine. Suffice Just any harm or hatred of a cell that carries life until her stage where he does not have the right to serve. EBE end. ALeLuja oo!

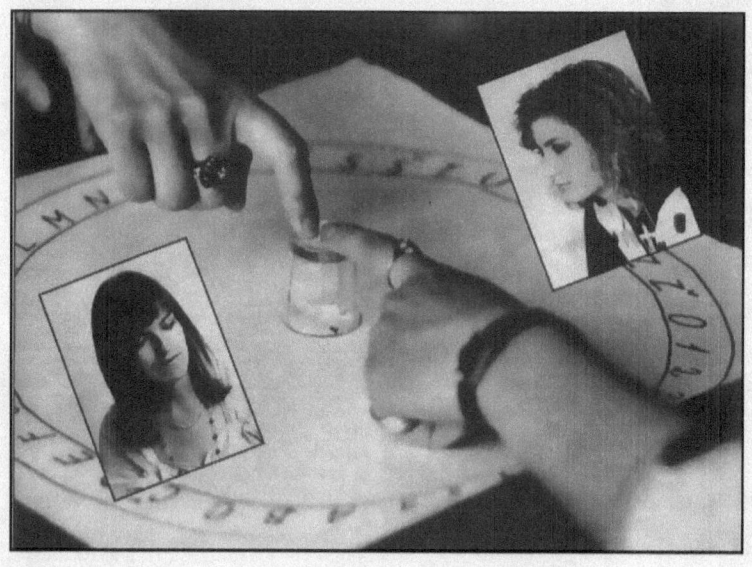

Your Leaders
7 September 2014

Aleluja OO! I welcome you ! There are a lot of new things. As for you and us. Much has changed. and changing. You have a longer period of time partly in another dimension. Everything is on a different frequency. Both technology and human development. With you on the planet is a lot of new space laboratories, which monitor the dangerous currents that surround you. Everything is under. control. Your leaders want you to have degenerated. On the other genetics. It is against the will of the cosmic law. They make and produces a secret structure. Both biological and technological. Planes, ships, cars and everything that has some power. AREA 51 have secret plans for the future. It's already programmed. People change in another STRUCTURE THINKING. That is wrong. It was a different plan, a different basis. People do not like you. Much gossip. We know people's relationship to you. We perceive. It's much envy. You have to ignore it. Will envy, career, Earth. It has to spread MUZUNA. Country panic. Another development, the other dimension. The time will be for you to run even faster. The Law of the Cosmos. You must look after yourself. They knows about you everywhere. Attention! Can anyone exploit, but we'll do, even if things do not intervene between people. But you will protect you from us! ", And what politician Skromach?" Scheduled yes. Nothing is coincidence. As you say. But it will be such knowledge. Which will be somehow affected. Freedom. Also the paper to change. Policies change. He has chaos. Ilona knows. He likes them both. He has a lot of acquaintances and friends. Career. Never mind that you have known politician. He is harmless to both you and us. Applying this life will be. The fact that you are experiencing is not a coincidence. But I think you also have to know what to talk about. Yes, soon you will see. Haha hachaku. It will be nice. . "And the older Mr. Simek?" I know he has no boundaries. He has heaviness in the head. Ivana   power drive . You are in another dimension, will pay seasons with you as before. Neither time. Everything has changed and is changing. It will take many people before you get used to this development. You two realize it. Others people   are restless and stampede. This is a big change. Next time we will communicate on a different basis of what will be new. It will be a lot of new things. Ivana is already tired. Will soon be communication. Have a good time ! ALeluja OO!

Civilization Collapsed

25 December 2014

ALeLuja OO ! I welcome you. There is no need to program. The thing is out there. It certainly everyone noticed. The earth is moved over a quarter more. It's a big shift. For your one year will be climate change system and the sky will be teeming with more colors. Will the wind. People will not be enough to hide. 2015 - until 2020. Then it will be very great change. Do not know your planeth . ,It will be a new dimension on another planet. It is so. People will be sad, but the money not lost. For people are important superficial values. Those who have the material advantages, and bursting with blindness to the outside world. Theye not see where they live, they trample upon what he thinks they have everything. But no are no nobles or important people. Civilization is terribly collapsed. This is due to the fact that you as a species been finalized. And that's programmed for this time, this time. It went wrong. And your planet is also experiencing chaos and twist. If mankind was to complete a higher level, so even better save mother earth. Evil brings evil. It will still be a lot of suffering to your country. WE HAVE TO MONITOR YOUR Earth . We're currently observe, but it's focus dazzling . THE SUN glaring FORCE sharp shine too large above ground. WE do investigate. THE GLARE , glaring OF YOUR PLANET OF.,,,, It's like when I look into a candle that burns a large flame. We know you know. The force is applied. You have a different system on the planet.

You have the hustle and lack of time. This is noticeable to the change that occurs. Since 2012, EVERYWHERE it was in codes in the fields. SPACE, SUN, EARTH , CHANGES. In crop circles. All connected of a certain circle. IT'S DEVELOPMENT something else. ENTRY INTO UNKNOWN. ALWAYS CHANGING repeated for millennia for some time. This is stage . They people were other than now. Other times, but the revision cycle is similarly incomplete, this is changing. Will They disasters in the sky. We are watching it on the monitor. You may not live in fear. What is dangerous, so these changes on Earth. We do not know exactly what will happen, but we know what it will be. Will come a time when you do not know yourself. I think you know everything important. WARNING TO PEOPLE WHAT deceive, believe in yourself. You Have eyes, you see. You have Soul perceive. EBE will have to terminate communication.
We must moving from place and act. IT'S TIME. ALeluja OO!

Do Not Trust Doctors

26 December 2014

Start of contact for 1993. spiritualism method .Humanoid EBE - Planeta Elieljí - 12 dimenze -39 Light Years.My several occasions we see with our own eyes and Humanoid object. Call the police. At first we were afraid. We know even paralyze. We are not alone here! COMMUNICATIONS LIVE spiritualism method. 26.12.2014 Czechien: Start of contact for 1993. spiritualism method. Humanoid EBE - Planeta Elieljí - 12 dimenze -39 Light Years.My several occasions we see with our own eyes and Humanoid object. Call the police. At first we were afraid. We know even paralyze. We are not alone here! COMMUNICATIONS LIVE spiritualism method. 26.12.2014 Czechien :ALELUJA OO! Welcome! As always We're ready to events ready for action. Yesterday I told you that your planet, your sun. Your overall direction shows a different part of the universe. About a quarter away. Your time is refracted in the middle minute. Only minute there you know it In the second, nor You can not breathe Do not trust doctors. It is perhaps a solace, but not on health. Your planet is in another part of the circulation. . It has around a particle orbiting different planets. We follow a consequence of nanoseconds. Never trust a naive and sophisticated people who know something somewhere and read the doctrine. It's just a gimmick and trap people. Do not believe! Believe only themselves! You're wanted for some reason to come here and you already know why. Did you know that you trust only

themselves. With this objective, you came here. The planet is in disarray with another universe.
It collision - part of the spectrum. Ivana is our goal..She has us,We have it. He has JAMAJA. Our pride. Mini pause on camera ...
Aleluja - Haleluja

I do not know whether you're ready Ilona,but Ivana always. Nevermind things around. At you one way or another Either look will focus on one thing. Repair is not. Me, I do not take EBE requirements for human beings. Either you want to or not.Me, I do not take EBE requirements for human beings. Either you want to or not. The law is what it is. Only mankind has failed fetuses normal spectrum of the brain. We went out one plan. We know how to attack against humanity. We need to have operations outside already, otherwise civilization will end soon. Yesterday MONITOR PLANET - EARTH -SUN. It is a signal for us MONITOR. We noticed a large layer, like sand in the Sahara above the atmosphere of your Country CAREFUL..Watch out! You have to worry about. But do not fear death, yet come Why, I do not know, but your country knows what you have to do. Many of you have not hurt after a few years, although in history it was certainly better .. In 1899 .Yes, it occurred renewal that time. Antiquity -Task of past life =" CARMA". You have to be ready. Yes, he fears sort,any contact public. We have work in other dimensions. We have a job in another dimension. We do not have time for it. We do not like any signal in front of the structure, the structure of earth. CAMERA - LENS OPTICAL - SIGNAL

CONNECTIONS - - ESENCE- AGREEMENT - PACT -

outrage and derision It verges on poverty. Although Ivana knows these things from the past life, makes you say. Fixing things here, Ivana exam and you know it. Daily Those that put the center of the cells when you sleep. When Ivana sleep. Therefore, you're tired and you know it. Do not mock us, ridicule. We know that Ty Ivana being superhuman. And because these plans with you Ivana share. Otherwise we would not do it. You're a medium You do not have your level. Why we do not know. She was lost. You have only to listen to his life, being. Heart beat him as well as to share. It belongs to the soul. I Vanga Your dog knows very well that we are connected. She knows. Your dog is also a medium of yesteryear. She Vanga dog is human. When you speak, she understands and you know it. It's not even Chew i do not lie. Her scratching , rubbing unrelated to dialysis doctors. But with her Task of past life "carma". Nothing not help ,only her cool and positive mind, and especially her honesty. Your Dog Vanga She deserves it. Yeah, your mom mother the message you, why not do that, as before. He is angry. She was the mother of a boy who is no longer a mother .. She is with you again! Does such a fate. It will be more want to go back into the body, but can only 2x. That's too bad, but somehow impede that. In the material world, it was worse. She had thoughts of suicide She said doctors at the hospital that he wanted to die, suffocated. They gave it to her, without you even knowing it. Why? We do not know. She would like to tell you that hurt her in the hospital. It worried her wounds, shock into the renal center. Where the tissue is the most sensitive.

And the human brain in the flaccid state has dementedly and mentally. Then the soul has a choice. She wanted to defend herself, but she could not say. I have EBE - ALTAKA. I'm not a ghost. I EBE, as always. I know that I do not believe many, many people. Our boats are daily on the horizon, but few who sees them. Try today, when you go to sleep listening to the sound of thin air EBE ashamed in front of the monitor. I'm sorry, I'm falling. I do not BAZILIKA map.frog, I guess. Love, Ivana, You were and are medium. People in former times you much admired Thee They had with you some faith. You're knew how to do what others could not, but so far it got. Just got blocked Many people are afraid to look into those eyes, because they have something others do not. And the people see them in black and white or blurry. Because these are the very people who are afraid to bojí.Co. Are afraid and believe. Your planet depends on certain stages. Certain phases. You either get zero phase or VAKUÁL PHASE. It is at a certain stage of the development process in the initial phase. It's like a race when you choose the path to a process of cell development. Each cell leads to the center and this is the same principal. Oh, and you want to know more? So let's be clear. We are certainly committed to you, you just need to fold the cover in which you live. And we will partly help. Just wish .. More are unable to communicate, Because Ivana is developing organisms in another dimension. Otherwise stabilized at certain frequencies. EBE must go instead. Important things to come. ALELUJA OO ! Hallelujah oo!! Welcome to a new year begrudge anything better. Yes, believe me! I know why. People wishing well and soul goes its way. Hallelujah OO !!! Thank EBE!

Ilona and Ivana with a friend

Two Suns in 2017

6 January 2015

(No believe in fairytales) -   Message from Ebe OLie , 12 dimension - telepathic recipient Ivana Podhrázská and Ilona Podhrázská ALeluja oo !   Welcome. Laughter heals, yes. If laughter is good. But now to the point. Your planet is shifting. It's serious. ZON as a dimension. EBE is infinite in the entire sphere of realism. Love is just look at the people. It has no physical meaning, but only mental ... Mankind has its drawbacks untidy formation of DNA. Cataclysm YOU EXPECT IN CURRENT RABID QUICKLY . It is not advisable to get into run things like politics. You have a body and soul to perfection and not at it's mired in selfishness and war unfamiliar to you. It's like a utopia. You've had one SIN on the planet as humanity. You know, we   watching you as humankind every moment. Nobody escapes us. You're DEVELOPMENT STAGE CYCLE. OUR BASE PEOPLE NOT KNOW. YOUR SPACE IS SUSPENDED OVER GOVERNMENT other universes. YOU are people for us MORE TEST STUDY IN LARGER PRINCIPLE on a priority basis. Your universe experienced a change. People were panicking at that time, was a big panic. That was in 1789. People have experienced some kind of discharge to the sun. it was averted synchronization Space , which also affected NOW YOUR TIME. I do not want to share with you the identity of the time in which you live. It is for us INTRODIMENSIONAL, MULTIDIMENSIONAL .It is blocking.

WE ARE BASED identity screening of cells, tissues humanity which our ancestors never finished and never does not happen to finished it. It is your purpose here to complete according to your liking. It is no destroy. Sun is changing. Do not worry, Ivana end of the world it still is. Interesting previously was year 1878. It was the beginning of the end, only. But everything comes at the right time again the sky . Will be seen in 2017, two suns. It will be a big storm, but humankind not to retreat . Do you have the time. November 2017 is realistic. EBE knows. We have here task . we're dealing with souls and with you . Until they not find their body , who either refuses or income. It depends on them. They have another dimension . But still this dimension covers the planet Earth. Therefore, they can be with you. Or at your place . Do not be afraid of them. We humanoids perceive souls of the dead. They, when they see you, they can not adapt to a different structure. They want to pull in you. You feel it and not feel. They will no accomplish until it no completes a his body. They usually have a new body soon. The main chief talk to them. He does not like us. We are for weak souls, apparently. They have their own, we have our own. I'm no humorous, but the soul of you people take power. I can tell you. Just when you think on them . Pause, I've got chaos. EBE is love "PALAVA - CHALAVA -MANOTA." This is our signal speaking to our ship, to keep current signal, that now time not lose , right now the monumental signal . Which is now a connected certain part for your dimension , atmosphere . The atmosphere is sort of signal into another dimension. It's like pseudonym "RACHAKA CAKO." Effort , respect . MALUZA HANALA MALAKA . Beyond what's to come you will see. But as I said at the beginning. The end is not yet. Attention, everything changes and the change will be. Because now falls often glass for our communication is a problem psyche Ivana. I'm from another planet . I'M THE DIGNITY. PEOPLE THAT does not believe but it's not our problem. Our problem is in the stars, not in people. People are numb, not participating with us to join. People not considered Mother Earth as their mother. You - people - Earth .Ebe for you people i will tell , through being Ivana one thing .

Why you do not like and why do we kill ? After all it is not needed. Consider that you do not believe us. Keep your MENTAL BALANCE. Similarly, everyone feels something else. And hypocrisy at you arose from a misunderstanding. Must you have the will to live! The time is quick . The universe is different towards you and to your planet. " MAMA KAKAJA LEJOPAPA." It's formation, creation , I say: Forget about Adam and Eve! That was just sabotage. It was a process, like a hologram. They were not in reality. You do not believe everything you since childhood has been narrated. Fanciful fairy tale had more sense of fun and history. They knew in the palaces, who lives and does not live in the world. Already in history in the Palace knew that Adam and Eve are figurines for creation . As a man and woman is the figure for multiply, and you do not be humiliated because of it. It's so much love in the world and it is hidden. EBE wanted to tell the truth.   You have to Earth fatal consequences of the misunderstandings how was when created . Everybody here create , nobody can imagine . Even if it is our work. God we all . Do not underestimate your life, nobody for you will have to pay for your life . That is all. EBE - OLIE end. I have my plan: Multidimensional measurements over your Earth . Aleluja oo!   ....

Soul

27 January 2015

Aleluja oo ! ! I Welcome you here! You have your certain territory. We cosmic watching you even beneath the ionosphere. From space, the ionosphere is visible and we are sometimes underneath, down beneath the ionosphere, because the ionosphere absorbs our source research. We're now below the ionosphere. We watch. That's our job. We monitor the adequacy of oxygen, which is related to the ionosphere. It's like the floor - in the gravitational degree. OXYGEN for you will wane. About We wrote about it earlier. We just want to keep the situation under control. YOU ARE UNIQUE PLANET. People are confused. PRINCIPLE OF UNIVERSE. So it was given. Neither sage, seer with you is not here your Earth, but from other worlds, yes. Whoever it was that all creation. It has been a lot of changes at once you on the planet, especially atmospheric changes . It's already past but repeats it. It's a cycle. It has been a lot of changes. He began to wind changed the climate. Was black sky, gray atmosphere. The lack of oxygen. People were confined in despair that took place in the sky climate atmosphere. Only one beam from a dark sky. It was like hell. People had to have an artificial connection that it did not kill them. They had to wear on their hands, as people withdrew her hands to the sky. Was Hands up to sky. Everything is connected, it's one big REGULATIONS. It's a stage "leaf. GAIA. It is so . It's a stage "leaf. GAIA. It is so .

You're at that time lived, too, were on other planets. Every soul bring something another to the next life on your country . I know that your scientists can not tell the truth. That around your planet are other planets, which you have previously not seen .They are like the nebula. Moon - Sun is in a different position. The fact should advise you of the source. But your source deceiving mankind, because of panic. Ivana medium is already from past lives. So-called ghosts. Soul of the universe are one and the same in a friendly dimension as well with ours. Everything is connected.

EVERYTHING IS ONE CABLE UNIVERSE . It is regulations system We just want to tell convey that we give to helping hand. People can not spit on us . Make impossible our dignity . Ivana, we know that your life is in the rapid execution of the material value as money. You are tired. I know that some do not like it when we emphasize in this communication, spirituality, the soul of the deceased and everything connected with it, but it belongs to everything what all surrounds . Everything's association , a large circle. Also circles in the fields. Circle - this is the basis. The universe is circular. It's a cycle. GAIA MAKA. Some people it know.   He knows about us. We're here and elsewhere. And other planets of interest to us.   Many planets, many dimensions and soul. People, when they die, their energy remains in the houses in which they lived. The human brain it feel . Soul already   has already prepared set for the body . It is a leap , jump   into another dimension . when a human dies, will remain part of energy and next the part energy require for

new   the body . It is an instant process . Many times it can undertake . and when the soul does not want to go on alone . So has the choice. In order to understand this, so I write about it . Ebe must end. Ivana decreasing energy. EBE - OLIE i am glad . See you soon. Just lift the head, do not look down at the ground.   The sky's the beauty of it all.
ALELUJA   OO !

24 April 2015

Communication with extraterrestrial EBE - Humanoid Eliel from the planet, Which is 12 dimensions. My sister Ivana is a medium and communicate from 1993. Communication 24.april 2015.: ALELUJA OO ! Welcome! We have a lot of new research in the area of bio planetary action. Bio planetary field is little activity in the war Suffocate us. There are a lot of poison in the air. the air is poison .. losing the sphere, as complex. we programming a lot of particles in bio sphere. It is the sphere with oxygen and atmospheric balance. On Earth, there is a strong organization intends That's destroy the atmosphere and ionosphere, and All That is Important for life. All realms we measure, we Investigate. We Observe, monitor spherical waves in a vacuum. It is the process of our work, Which We currently follow. Everything starts to lose the his balance. Your Entire solar system, as you type in the textbook. The solar system is primitive name, but your name.System , as droughts, is based on a different development. There are many other planets That have not entered to your history. It's Another whole solar system. The sun is fading CENTRÁL .. all the poisons in the universe. If this will continue, it will be a really serious situation Cosmos. They are covert operations about Which we know only we and others outside of your country. It is a plan on a large scale. The sun wanes, not much saturation for your country. It targeted.otherwise, or else would you sun burned. I already telling .. The sun will last a little. Year after year, Will weak sun. On the whole earth. everywhere in all states it will feel ..

sometimes less, some more .. We have it Monitored. Do not worry, you're not, but it will Slowly. We are afraid to get close to your atmosphere. We monitor it from afar, We must be a Certain distance. Could a lot of energy protesting for our boat. Could we harmful .Not afraid soon to come. It's the law. We control the sun, watching the sun .. Earth is a small bug in the Cosmos. Will be 2 sun up to sun Reaches your minimum, minimum, and a second sun Appear to be less clear, will dim. It will be for a long time, but you year on planet Earth too fast running out. Speeded up your run. Quick Time .. Earth's axis is slanted, crooked, Uneven, tilted and bounce out of orbit is tilted ITS axis .. It is written the law in the cosmos, as in your Bible. It's already Happened, it was once. Cosmos law is not only That it will drop, shedding spherical mass. That will Decrease spherical mass. Each sphere to your country for the development of life Cosmos law is really.   it, but the Bible" Scripure" has a lot of questions, the Bible,scripture   is true. . Another Bible is from dimenze.Cosmos, the universe is the law is valid for all dimensions, planets, cosmos and beings. We are scholars of His scholars .People   Their religion and Their Duties. It's a learning materiál. Only teaching people to live Somehow. Other laws are taboo for people. It is sad shame That many people are at a computer. Computers interact, manipulate people's minds. Computers have a negative effect on the brains of people. Brain cells. State, the situation is drought That for us, as extraterrestrial beings Mentioned in miniature computers That are on different videos. A lot of it is a fallacy, having . unfortunate evolutioní's all true, what is Written about us, and videos in your computer. It is Constructed from a film programmer., Who likes stunt .. It is to see if it's true or not. You can not believe everything. You have to believe what's out there in the sky, you see only your eyes and your Consciousness to perceive. Trust yourself .. It's anywhere you like everything to verify. Our alien ship reported a record signal, we have a signal,   defect . .. We fly close, close your atmosphere. We have not Had an accident, break anything. We soon   danger Time signal indicates our boat. We Amaranthus Could Achieve a multi vibrational energy low. Seen, studied when, your ambiance.Whether weak,frail..People feel great fatigue and pain back, most of Which Operate on the bone in the body, the legs ... your bones, bone and spine Affects. When someone is tired, my body is there for people fatigue and activity very many Widespread expert program. We know the impact of the weather on one or a ray of sunshine. You people have computers really for us primitive. People having primitive Computers . Computers   with us is a true need, your not a true act ill people. Deceive mankind. It's not for you to a Higher Evolution ..U you think of computer smart people, educated but not completed the technology Principles for good and true things. Only 60% is true in your computer. We have more information. We got back, now close to the atmosphere. We'll have to move away. We have gone out. We must turn to open   all, everything   blocks, MSS block, Which have the task to move away .. I do not know. We have a problem, block,. State of the atmosphere, the atmosphere Acts Causing on our boat. And we answer to the atmosphere .. It really without power. EBE is now too complex situation, big complication, the situation with others in Lodi. We are only 3 Humanoids in   alien ship   . We must ending communication and we will try to fix . We must situation, our state. Ivana, we check to you, watching, do not worry Your health status known, it is not dangerous. You have to pay attention to your health. We can not help in what you do and how at the moment, Ivano live in your body on earth. It's your choice. WE CAN only help Those That will follow you inside, but it's not easy. We follow Your energy .We   must end. ALELUJA OO !

Ahoj! EBE   - OLie - Message:

2 September 2015

Unaltered info send full transcript of the communication from the
humanoid Ebe Olie on 2 September 2015. Whether you believe or do not believe. It's up to you whether you believe. !
We do not force anyone to believe us. Also, of which we have no funding, not earning on   it,, but we devote our time. It gives us some understanding and 100%, we know that we are not alone in the Universe. So the judgment make yourself .

Our ship SERVES AS ONE HUGE LABORATORY

Ahoj ILona ...Message from EBE : 2.9. 2015- ALELUJA OO I welcome you! We have a new reading about your country. The monitor shows us the curtain that goes through your osou. Curtain is shining brightly. From every angle develops many images. Our ship SERVES AS ONE HUGE LABORATORY. It is isolated. We do the images. Research laboratories and we are there. As a probe without you. But that our works on the basis of molecules of carbon and iron radiates neutrons. Therefore, it is isolated at a certain angle for us. This people have a different technique, such as a toy without the kind of substance and the molecules of the various elements. Ship to dissolve melt from our many molecules of iron and carbon again on a different basis. That use. It's like our source of electricity and techniques used in this system. Can your scientists know about it. Your scientists

would like our basic resources, but he would not understand, they would only hurt. It's a dangerous thing Your leaders would like to seize anything that they do not fall. Axis sphere and other galactic affairs. Your energy maneuvers climate makes does not contribute for a perfect life. We know you are ,
We scanning ... Everything is on the other developments that had gone wrong. Your I they. Your powerfull leaders of the world ,they have People are, as the experimental mice in a cage .... think only about money. Intelligence has turned into a fake plan. Who rules the earth. Most people are asleep, they succumbed to manipulate their minds. Second Sun is but a curtain. We know who's causing. But it's there. Someone has the opportunity to see through the curtain. What on earth are you on, it conceals, but what is or is not, so he sees it as faith. That is, it will be something unprecedented, and nothing ... That people believe, but the truth is hidden. The truth is you secret. But it is obvious. Ilona, Ivano, people will you still block in comparison with other people. We know that these sensations you open up a lot more than before. Yes i Cosmic Gate opens you more. Perceives is the one who is strong, who has sufficient intelligence and development. He created the universe. You're younger. We have you formed. You have originated differently than we do. And they had the technology to the development of yours had the necessary foundation. Otherwise you would have been similar, as embryos or aquatics. Which would not be possible planet Earth. You are the only one such planet. We did it previously inhabited, too. But now we have other laws of

space. Inhabiting her more species than just us. We stayed before on planet Earth, as it is now sometimes above the surface of yours. We're perfect for that of our planet and it is sufficient for our development. You're the youngest in the Universe! Before you were created, we already had long since advanced technology. So then for us it was a breeze to tighten your structure to a higher order. Take care! Ivana and Ilona, you're under great surveillance of certain groups of people that you know i do not know, you know and you do not know. Under our supervision least you are safe. and under the supervision of other people are in danger. Do you have a shaman, suddenly is love, career, love. Ebe - Olie is on the principle of a weak moment now. We have a lot of work. For you say work "Slavery." For us it works:"Energy" Energy in connection with the energy of different elements. Ebe must end. Question: : Ebe, what do you think of the Anunnaki? They Anunnaki are a number of laws. We are with them we abrogate laws. They have different challenges than we do. They will probably find it knows about you. Already i will ending. ALELUJA OO!

22 November 2015

Communication with Ebe - OLie - November 22, 2015 - ALELUJA OO! I welcome you! Careers end of the beginning. The earth rotates in the other direction, people do not observe. . Everything has gained momentum, and the other direction. So it once was. Chaos in the universe. . We are watching what you experience on Earth is happening. On all sides there is danger. Humanity is under great pressure. Two Sun are, but this one is still far away and hidden. Rarely seen. It's the law. . Up to 2 sun One is larger and the other smaller. To come closer to each other, so there is a break. So it happened and will happen. Are you in danger. People are becoming angry at themselves because they feel fear of the unknown. . Everyone is under the program. . Beware of people. . We perceive, Ivana and Ilona, what's happening around you. You're ours. Even in other dimensions they know. People talk about you a lot of strange things. You have in your house evil. We feel it. Do you have someone new in the house strange energy. We can not yet come to power, which will handle. Energy is able to achieve a lot of pressure. ..ILona Must have pride. Ilona's mastery of the whole. Try to drive off, what you have at home. It's in front of us isolated. He is interested in you enough people, watch it! You have to protect, like the Earth to be protected. . I know, Ilona think that today we do not disclose something according to their liking that flirts unusual statement. Every our message makes sense. Ilona must all write about us, it's dangerous. I need to discuss your condition even though it usually do not do. Something has changed with you. Not enough to be surprised. Someone wants to take, seize. Ebe is outside the program to find out who wants to seize. Q / What do you think about Ebe Anunnaki? / Bright Lights. You know. Many people are familiar. Country know. Anunnaki once ruled your planet. Entities are huge and have a great voice, as you do when talking robot, powerful, deep voice. We can not in an emergency all communicate. What they convey is to be designed especially for you, and if the people you know and who are safe, but do not know it. Is information that we would like to tell, but we know that we must for security reasons. Entity more about you know Ivana and ILona, And to recite, one of you tries. Not seen yet, it's just energy. That someone is playing with the energy of people and can be isolated. This is a novelty for us. But we have not come across yet, but we encounter. We'll get to why various stages to determine who it is. We plug in the frequency of various types and isolate . I will not say more. You have to have pride. Forgive me that today, so I convey. You do not know who's interested in you now ... It's going to seriousness. Cautiously write. I told you what I had. EBE yes chaos must end. Think to yourself and to the animals. So, next time we will again communicate. ALELUJA OO!

1 December 2015

Communication with humanoid - EBE named Olie
Communication. Spritism.. method. Planet Elieljí . 12 dimension ... Live contact.
Sister Ivana and Ilona Podhrázská. Ivana is medium. Czech Republic.
Today is December 1, 2015, and we decided to make this communication with alien EBE.Humanoid EBE - OLIE , 12 Dimension Communication has been ongoing since 1993 We decided to make this communication for people to know how to communicate. The humanoid EBE name Olie. Ebe, we welcome you! ALELUJA OO ! I welcome you ! You have a lot of material in your system Which is hidden in the hydrosphere. System absorbs the organic structure in his heart Is it unknown for people Do you have a lot for people hidden and unknown little explored Your planet is your humanity unexplored. Your planet we know very well We've got your planet, as icing on the cake as you people say Should scientists and leaders of all countries On your planet join forces and brain centers and try to discover all that is needed. Humanity lives only material But the spiritual equation are hidden as all unexplored Target ages for people not mature, squeamish Life on your planet is sad for the people , when laws are not appropriate equation in your human boxes We are currently investigating radiation above the atmosphere over Europe Somehow adapt quickly stopped the system is given. spin out of control Everything we recorded we see brightly shining signal What is just a

block or shifted
in your atmosphere We are shooting everything, but also we have the hiding place It is our protection even so, we are watchingeven so, we are watching we have a different structure our structure is hard adapt to your structure as for you to another dimension us for our dimension Our laboratory is different than your Your Laboratory investigates primitive Our Laboratory investigates all substantial and Cosmic law given on a different basis of development molecular center You people make up the buildings we not have a buildings, but we have under buildings which he took time at a time when the track has changed in our dimension But we have othe houses" Bunkovka" Say " Bunkovka" , not houses t's protected and isolated for us you can not destroy it. Indestructible You can not imagine what it's" Bunkovka" house for our life , existence It's similar , as for you on your Earth like tinfoil , silver paper. We have this , like tinfoil For us it is a protection against dangers It Protects our bodies , in fact our lives once in history we had houses but they looked like chimneys, like towers but other than yours people, your house will never protect against the danger of any. not having protection protection against cosmic catastrophe Never will not protect you from the war system Already that it is your on the planet and your way ... For us do not like And it is the more of these dimensions, which do not like it We war not we not leads. No war We lead the cosmic peace
it is not a universal law of war point People themselves not tolerate, we do not understand why Why must develop weapons on Earth? we do not understand t is unnecessary.

You are the only planet of war In all , throughout planetary block This is block , at us Your are so tiny, small planet and so many war but will the cosmic tidiness, cleanness your it compares the understanding Comes the time to be awakened humanity from those harmful things Question: Ebe, Can you describe your body? our body we have previously described Truth . Long ago We have a different structure but for security it is not possible to know more We have another dimension. we're different It is clear We Getting to you ,so as given and need, requirement we are here to jump so it do recalculate At us calculations is not. It is infinity It is only a space. And no time We have large and small. we are different We do not have blood. We have white liquid, fluid For you, it's like blood. And we do not have veins We have only tube and nonliving The tube We circulating white liquid otherwise we based biology Yes, that's us. But otherwise , than you humans also we live Not so close to you, our dimension is not your dimension. at it to ensure that you know too much about us We are afraid of humanity We know about you all the information But we the people not hurt It would we have been done a long time ago we do not have guns , weapon People are scared of us. and we are afraid of you people too and you people have guns Yes, it's too much When someone will have the opportunity and wants so we reveal. Crop up. Yet but not so promptly It still prefer isolation It's not easy interconnection of all beings in the cosmic order Yes. Ivana has great abilities And Ivana is tired We see that it is tired the time has come order, regulations we like you we say that people really have no protection This is on our side warnings

OLie - Ebe have to keep going over the atmosphere We were close Our Ship gives a signal audible and foggy that it's time to do another task We send word the people what was needed But more important information they are for Ivana and ILona extremely dangerous so please understanding I say it here Here u devices, apparatus , camera Ivana has little energy And already we must... End Keep in the hearts of peace ! ALELUJA OO , yes Aleluja oo And Olie I am not he or she, as it is at you We do not formation, as people ALELUJA OO! EBE we thank you! We will communicate again ? Yes ! ALELUJA OO !

We have tribes

3 January 2016

Komunikace s Ebe- OLie   Communication with Ebe- OLie 3.Januar .2016 : ALeluja   oo! You are in the   change . The change   point came. Energy are evident in every part of your body. You have the ceremony in a large resolution associated with particles of the entire process to biological and chemical level. Do you have something to do with higher sférama. We signal on time. Your cameras do not interest us. Your systems go out of our members who are in Ship. We lined up in groups as an auxiliary system in the trunk Elieljí . We have tribes. We're dealing with the elite tribes from other realms that are in its way to us tactless. Your earthly people are too different groups, but for a stunt. We are part of the universal soul. . We have a clear energy. You people have a severe energy as it combines hard-s poisons . These are poisons, they get to the stage of decay energies and bad vision. A hidden value of mind. We will scale. People are just figures, trying to control the entire system. For you on Earth is to use a system of patent systems of the Earth in secret. Question: What Contact CE-5: / These are groups we know about it. It is important that groups of people emit. It must be clear mind, energy and heart . When that someone in the group thought about money or sex and other ailments. We'll tell you one thing. We know very well that every person has a mind busy. People think materially. We see everything. Lazer us violates our management in   Ship . We still can not get closer, but soon it occurs. You people say, why more close. It's been many years since we have had to show. For you it is a different time jump. You want everything immediately. But this cosmic system still did not command. Ivana, your leg heals. You can work where you worked in a meatpacking plant. You're going elsewhere, because your foot, it was the beginning. It had to happen. Meat and You, Ivana, it does not make you worked there. Those here's another task. There was a mighty Your "self" It affects your body. Eat meat you can, but you're not judged to work as a slave in a meatpacking plant. Do you have a career. It was a short circuit and will, if you're there to work. Otherwise, you Ivana healthy. Do not worry, your liver can see are clean, pink. Tests by doctors are unclear. Your doctors are not sages. All you carry your mind in your bodies and minds that you manage. , But does not control your destiny. What we were given was not affected. Ivana energy fades. For many people, you're special. We must go further. He calls us to our "ŽURAJA" There's a light with our system in your area. There are a lot of poison. . Ebe - Olie must end. We wish you: Live you continues to balanced! With respect to you Ebe - Olie say and no scrapes ,seamlessly , safely. no problems. ALeluja oo !

Questions From Six Researchers

21 February 2016

Original -Message : Spiritism Communication with Humanoid Ebe- OLie. with questions from six researchers: Tonny Topping, Phil Kava, Malcolm Robinson, Bill Rooke, Albert A Rosales, UFO author. EBE : 21. 2. 2016 -ALELUJA OO! I welcome you! It's crowded everywhere on Earth's changing energy. Seeing energy in the Universe over your atmosphere acts as a diaphragm in a matt cover. Energy can not be seen, but seeing it in your device. It acts like living matter in a closed cell. Our signal sends rays of our ships often into your sphere. We examine your resources in climate. Currently we are watching this. Nothing is suitable for our work. Our ship is clamped into other dimensions. Your country is neutral at this moment. There was also a change occurs, but so far is neutral. Question for Ebe: from Tony Topping: "Who is the universal architect" ? Universal Architect is a powerful but not most powerful. It is the perfection of the system that controls the planet Earth and the nearest star system. Tony Topping we have no the record, but it is a researcher in a certain direction, but the identification is not linked to us. "What Phil Kava"? ? - Yes, but Tony have another contact. Has a different civilization does not know us. Not us a message too Phil kava is based on scientific research. It is the identical with more people who are interestee in alien trophies. " Question by Malcolm Robinson: "Why were by abductees two men in Scotland on the road and 70 in 1992. What was the purpose?" ? -Many people are abductee from different countries to your country. People are abductee in research with other dimensions. They signed a contract that is identical to other dimensions . . More people abducted .. Those two guys, are whether tall and without overweight and if you have short hair. If it's them, so they are used for research with beings from another dimension. It What investigate , I can not reveal. It is a contract,, If they were different in appearance, so they were only for the quality of the sample and development. They were abducees by another party. It's more creatures that take people unawares for different targets. Either for research or for explore It is more directions. "Bill Rooke " give question: How many light beings is in contact with Bill? "? - York City know! We quite often there are scanning from the atmosphere. Happening there Magnetic Resonance often enough. It is true that in York there are many people who are in contact with other civilizations. beings of light , may be more of them ... But there is always the only commander who leads a group of beings. It has a large group, but is also Commander in Chief of light beings. And they are hiding in different dimensions. No comes directly from another planet, from any planetary system. They are different and elsewhere. It's not civilized people or beings. They are transparent, translucent , like clear water. They no are see but a little squeaky emit a tone. Whistling. catcalls. And can someone people heard it. Bill yes, he can register. Can heard. He may be in direct contact with the commander, chief. " Question " Albert S Rosales :"Who will be the American president?" ? In

America, there will be a long time leader of President Obama. Previously selected None. Waiting on a higher power, which is still in development. So far it is the hidden development leader. No leader has no such power, as is the Cosmic Law. Therefore, it should be developed, evolution. Someone who is so mighty power to control the entire planet Earth. For your planet Earth should rule 1 leader. But God is universe. Someone like God, But there comes a time when the trend is born. It's not time yet. Triggering factor is the universe that is changing, has changed and the change will. You, the people and the country you are currently neutral. What is happening is out of your space in the Universe. Planets are living, changing their regulations. approaching the planet is to see. When it tightly close to the sun so there will is a big break stage. Little who sees it. Actually, defacto it is hidden to humans. I said last time about 2 sun. People it will name so i named. " Question from Ufo author: - " What is your current position ? We are from another galaxy and ,, various dimensions. We face the . I investigate, we see everything that happens in dimensions. We examine the universe more and more things than people. We already know people from time immemorial. "Another question from UFO author: What is the nature of your existence, you understand how the human mind, what are your senses ? How many people are you in contact?" Our existence is built on a different basis of development. WE HAVE ANOTHER IN PRINCIPLE mass in biological CENTRE OF CELLS. Typical of development. Boddies Organs we have stunted but otherwise we creation as a whole in the universe. We're not negative.We no manage the mind control . We no think like you, people! Definitely not. We follow the different intelligences. We are able to connect with people, but those are really little, same as Ivana . We are only moderately lenient and advanced intelligence. Lenient towards people , but not all . Few people we believe that all mean well and seriousl people . We have no senses. Our energy gives us direction such that it was serious, intelligent, and so we were not making fun of other beings or of our own species. We all take dignity and severity. ,People   very triffle when the universe offends or falsifying records when his identity. deceives the body soul and spirit deceives the body. WE have   essence, which gave us the creator of the universe. It is the essence of living. ATTRIBUTE message. Messages is given so as to be visible from a large part of the atmosphere. You have a lot of them. The messages of the grain in the fields. But people are not dignified . People so considering that it is a creation message is from you people. that people do and that message depreciates. It is without importance. It sees human sense, such as the eye, it is very little. It's like a grain in a field of poppies. There   is so much that the human eye   can not see. It is like the sea, as if to bring together all the oceans. We not have touch, sight, smell and all your cells, you get into these senses we do not have. We we follow ofter mosaic cells, inexplicable for people. we are different. We see from the inside. Our eyes are only, like a flashlight. We feel from the inside on a different base. The moon is KACHAKA ŽAKAMA. Moon rotates in another direction, But leaders do not say for people. NASA silent ... People should look more to heaven than to look at the ground, To have some here pettiness. Material thing....for which kills. Absurd. Yes, Ebe end. We are no more time. we climb higher and higher. Again sometimes next time! ALELUJA   OO !

3 April 2016

Aleluja oo! I welcome you ! It's a lot of changes. In space and other systems. The understanding of the universe is a lot STRONG magnetic energy all around you. Nature undulates in various areas. Ripple is a strong current wave radiation into space. Mutation of nature is on the verge of new developments in Space systems whole. Man is a part of it. There are a lot of changes in waves. Our ship logs from a height of many miles your country. The last time we see your Earth in fog. The country is in a fog and in an inappropriate discrepancy. Waves and mutation in nature is guided by this. Heavenly planet is always ready for what's supposed to happen. I've already telling. " SPACE THEATER " and its history repeats itself. Only these changes to nature are inconsistent with history. Everything must be ready every individual. Few individuals know what will happen. The political situation is for you also inconsistent. Should not it be the way it is. Went wrong with politics in many scenes. Sad scenario, but it is a remedy, we know about it. No good plan is not in politics. COMING NEW SYSTEM .O that I have also telling in previous communications. I guess " VIACHA". It's the circle of life earthlings. Each individual must take it you're not as Earthlings, you Spacers. Just as we and other dimensions are part of the Universe. The universe is a unity and everything " RAMANA " chaos. Country chaos. We know what people in your area you plan. A sensor in this behavior, attention give to people. Be alert and be unique, then you guys will respect. They do not say anything good about you. We see it, are powerful energy . All records to our device, such a warning. We now have a SIGNAL FROM OUR DOME THAT CURRENTLY PROVIDE MAGNETIC INTERFERENCE , because we have moved a few miles away, and we have to be closer. SPARK is strong. . WAY OVER YOUR PLANET do not feel safe for a longer TIME, YOUR TIME. IN OUR TIME THERE. This is just information, time. / Ebe, write something professional? / - I pressure.

Vocational emphasis in this communication may be dangerous for you if you are not misused for some leaders, it has no answer. Leaders they know a lot about us and you trying to manipulation. It's dangerous, watch out for it. Are you at a point where the safety of dangerous .You will escalate in danger. WE HAVE BOAT disruption. If we ALL AND OTHER STRUCTURES IN THE UNIVERSE WOULD YOU MAKE info A dignified, professional statement, so it had we would have long sience PUBLICLY IN A great discovery, I. It has yet to be, yet it still does not but it will. Chemistry, we communicate and we will once again be repeated. Whatever your scientists put questions. I do not know what you would like to know on what basis. Today we told YOU what we wanted. TODAY ARE MORE OF US IN OUR SHIP. We are five humanoids in our Ship. We have a country career. Chemistry as such you take it differently than we do. Not force us, just for someone to target. Interesting, yes. You may not know about the world. It's dangerous. LOVE AND THE GODS ... COUNTRY CAREER. Suddenly career. It is not over yet. You must not be hasty. CHAOS IN THE AIR. / And what Anunnaki? / - Anunnaki are on the line. They are hidden. It is not yet the right time, but humanity will soon introduce Anunnaki. They are afraid as we are. Country surges yes. Olie will be communicated later. TODAY IS NOT SUITABLE ENERGY from several sides. Olie ending, yes. A L L E U J A. O O !

19 April 2016

Aleluja   oo ! Welcome! The change is whole   a planetary system. SOLAR SYSTEM YOUR COUNTRY SOON engulf strong energy from the cosmos. The Law of the Universe. People are like robots controlled system to manipulation.
. ... Particles that are moving in a rotating time. They ARE dimension that absorbs light on your planet. . Your sun fading, losing power. Gradually, your country is moving in larger frequencies.   There will be many poison and disasters before you expose a new sun. New Sun will be near to normal Sun . You say NIBIRU, but it's like your Sun. There is none   life. Nothing, just wandering and wandering object that has its track as all the planets.   Galaxy has experienced other disasters. People are manipulated and still are an endangered energy.   And we perceive VIBRATIONS A POSITION Earth's core. CHANGES IN THE EARTH   FROM THE NEW SUN THREAT . Chaos in the clouds. The rain is full of poison. .. SUPER APPARATUS FOR REMOVAL OF SAMPLES TO   OUR MONITOR   SHOWS   SCARY , WHAT IN THE AIR YOU   BREATHE . Atmosphere is weak . Operations are set at the following gases which   in the atmosphere missing . Argument with humans. I think source to be scientific equipment and that transports many people different direction false. Program " MAKAKA " Light program in our blocking apparatus . Already our ship is repaired. Blocking is less ." GAMAKA SHAMAN MAKUBA MATAJA". Yes, I convey our speech when the signal drops out between me and Ivana. We see you in Ivano change. Do you have a

lot of light particles in the brain. When will space events, so we have with humanity connect. People always say: "It's still around, we so as prove, that we are . and yet nothing. But will it comes to suddenly as it had come suddenly crash   in the earth's core. EVERYTHING IS GOING ON THE BASSED ON THE MOMENT. IS SPACE CHANGE THE SYSTEM PROGRAMING AND REPEATS . May change the dimension which interpenetration as in your system and in another system . On your Earth are crucial three things. which are bad. It is human shortage in understanding , abuse of Love fullness and harming Mother Earth. Her kingdom, which is called the solar system and its Galaxy. I have . And you our " NATIVA " Your Earth we have control . And my Lower Sub-Commanders they guarding all signals . Radio - signals from our Ship . Our Ship is a control. Our Ship is a connection with our task . Yes, we will communicate again sometime. OLie will have to end . Aleluja oo !

The Hot Sun

29 May 2016

Greatly quick writing contact , communication with Ebe - OLie : Aleluja oo ! The sun is hot and the whole cosmos. All particles are in constant pressure. The particles in the universe hates with the melt particles, which are opposed. Everything is on a particular system. Energy particles form a chemical reaction, which flows into all layers of oxygen. Oxygen is unique in the vacuum of space. It is part of space in which a vacuum. Chamber in which the oxygen is held and can not go into the space. Space is at a basis other than vacuum. The vacuum is loaded with air and space is loaded with oxygen. Nitrogen absorbs everything in the vacuum of space. Nitrogen is a part of the molecules that form the neutralizing energy source of the cell wall. It is the evolution of the cosmic principle. We're now over your atmosphere. We have a lot of new spherical molecules that during our trip we were to take into our computer. We have a lot of work to form a cell to another educational system, which will be justified to the human damage. The human nucleus and brain is large damaged. It is caused by a chemical process, which does not protect human beings, but it hurts them. We have a strong pressure over your atmosphere. Our computer shows strong waves magnetic storms that manipulate our ship. We must hold our source, which is protected from the pressure of your atmosphere. Your atmosphere makes strikes more than ever before. Toxins are a test for human beings. Second Sun is hidden intentionally. Only someone somewhere sees second sun, much as when you catch a glimpse of us. Also in a certain angle. Second Sun is more certain angle. It is more an angle of about 5 angles , which can be located a second sun. Your leaders can not handle. He does not know why the sun is hidden and sometimes not. They are trying to put more venom into ozone the sky was covered with more. Your leaders engage in multiple spheres poisons . Breathe in poisons. It's a trap to test people. Chaos around the Earth and chaos in people. Everything is connected. Animals and nature are suffering, crying that superficial humiliation. Career test on humans. It is more alien beings among you . But not us. There are several others. They are more for you. People have feelings and not know what's going on. People perceive the world in another dimension. You have a dual world. Spiritual and physical. Suddenly you can not perceive. I guess we have a signal on the surface. Beware of evil creatures and people around you. Safety is the essence of life. Ebe the task. Ebe needs to higher realms. Soon we meet . See you soon . Nibiru is the second Sun from another dimension that revolve around an axis, and a few thousand years back to your country. Earthquake shock is in the Earth's core. It's a big shift magnetism. Earthquakes will be oftentimes . Ebe have to go, let us sail. We have a task that is important   because of   your leaders from around the world. Aleluja oo! "Ebe must end Aleluja oo !"

10 AUGUST 2016

ALeLuja oo! .. We're not, We're not like you people. We have another source of cellular energy centers. Your human cells of all Earthlings have fallen into the depths. JALAKA. The cosmos is not a part your learning. It is endless, but you people never see this part and you, and never understand it. The cosmos is based on a different structure that undulates at an angle on the principle of carbon, and all of the substances from which come from the Cosmos. Carbon is an important part of the element . An element that is not yet fully explored for the development of your country.

Note; Bucky Balls and other carbon based nano technology.

Your Country has teamed up with other parts of the planet and the disintegration of the Moon all caused the carbon, which at the time of creation was not present. Then came reintegration. The carbon part has changed and it is unexplored, because something has happened. Disintegration in which small planetary particles in the Cosmos is meant for a part of the element carbon to collapse. When there is a collapse of cells, the body is silent, dying. So it's the same in the Cosmos. We have our school, which is given. And you learn something from someone you use to know. Many things are hidden from common knowledge. Everything is a total fallacy. Why do parents raise children in such filth, who teach only falsehood. Fallacy is nothingness in which   people wade . You have gone through so many things. We are sorry . You are unformed, but the science that was mistakenly given to which is not related , That was the only purpose and destiny of those who are living in your physical system. Your part of the universe is calling for help! ! We have another part of the universe. The universe is the source of all power structures in the center of development. Your inclination of Earth, your angle of the Earth is tilted too much. You will see a different angle of the stars and other constellations. Everything is disturbed in your universe. It is triggered by manipulation, which has been going on for a long time now, and that makes maneuvers in your Cosmos very difficult. Mother Earth is part of it. Nothing will be visible above the other, as before, as I have already mentioned. Everything changed. The Moon and Sun know it. The sun is already dying down, we're sorry, but people are asleep. These people, the sons of man, they live in the material world have no clue. These people do not see, so these things are hidden, and will not feel that there is a shift. Ivana and civilized dimension Zon are our dear friends.They're watching you, Ivana. They are dignified, courteous and dangerous in the sense that it loses its dignity. They do not like me telling you these things. Only things that go their way . They are indiscriminate. They go for his goal. They are a dimension that is quite distant from your dimension and can mingle in a certain developmental spectrum. Which tracks information on when to run the main point of contact with you. They have only a few people, like us, only a few people . Ivana, theyfound you, because we've had sealed a pact! We, together with the civilized dimension Zon had to go through

the school, which applies to all living things in the universe.
Such as training. Ivana, You're here because of a certain goal.
Therefore, Your mother finally gave birth to you. Otherwise
you would not be here. That's how much you mother wanted
you and so we said that at this place you Ivana can be, But
you had the choice. You're away from us and from Zon
dimension. They have known you long ago. It's hard to
understand for you, but it is so . We listened to the prayers of
your mother and you're here in your beautiful body. You could
have been in another place. But it wasn't your destiny. We
are glad that things turned out this way, and we guard you!
You have a different system in the universe, which Has an
ending period. Which had four periods. You will have only 3
years 2 periods Time in another. People will talk, say. Talk
between the busy street. People will be talking about children
who were born and to whom a child, and who lives with
whom,   But people do not perceive. They will not see the
changes, abnormally large changes. It is a turning point and
that period is not, as intended, for mothers and children when
the world was normal. New children will be another kind . It
will bear genius, and a greater manipulation of blood, plasma
and whole center cells . Children, as it is, then adult children
will be robots. They will have a few years of a chip inside,
which will gradually change them into robots, and so the time
of robotics will come to pass. Everything is already done.
According to the ideas of ancient immemorial civilization,
which caused this law. We've brought you into the world and
have not finished the creation. We are in an ancient
civilization, we were like slaves, and that you have not

finished it, and now the cycle is repeating . Those children like robots will see as the Moon. and see the whole system together, abreast. None higher or above the other. The end of the world actually occurred, as for your civilization, which was born from other layers of being. We now have a plan. We need something to be corrected, the material, which is in the soil in the ground. We sail ... Next time we will communicate / Ebe, one question. Who supposed to win an election in America? / -   I now will not answer your questions.. / Ebe and who will give the chips to those children ?   / Ilona, I will not answer. It's a dangerous answer for you . We know that some people already knew more than he should know. Yes, be careful! As for you. Seriously , someone, people are going after you. You're in danger. Attention !   Ebe result , must end   . We have plans to correct the soil. We must go ... Now we are in the void of the Cosmos. Yes, soil, plants Ilona it is too much to ask. Soil, plants, specimens for research. We will examine the level of toxins in your soil to your country. Yes, you have many poisons with you on Earth .   Ebe must end. Aleluja oo ! "

Particles

30 August 2016

Spiritism method Communication with Humanoid EBE - OLie . Contactee wit medium Ivana Podhrázská and ILona Podhrázská from Czech republic : Aleluja oo! I welcome you each other . On Earth is decreasing energy point, which is connected with the cosmic convection . Flows to you particle , which loses on the strength of the atmosphere . Oxygen absorbs this particle . Particles is created on the principle of a single molecule in connection with the uranium , which forms a mini particles in the atmosphere and manages thereby operation activity effect Meta Gamma . Meta gamma - is a source cells for us prosperously for observation . We investigate the evolution of the universe on your Earth. It changed everything in your cosmic order. . Your universe is new and has a development of a new order.Your universe - your dimensions. The universe is all beings . . Other system of the universe. We have a different set of planets. Yes, you can see the planets in the evening and the new Galaxy, another grouping only of partly . When will an entirely new space system . So the of earth you will see the stars like the earth.So the earth will see the stars and the earth. Will be within reach . But doing so will be several light years distant . You are at a different angle. You are in the middle and coming fully turnover , turn Earth system. . Will other track, way. I know you talked about yesterday a special phenomenon in the sky. Yes, this is it. You will gradually seeing another galaxy, other systems on which your country has not ever accustomed . In this history is not repeated. It goes further ...   But 2 Sun , yes . This is repeated. Molecules control center in vacuum like trajectories the protons'. Proton is very important in the investigation and findings of all options for a particular development. Oxygen   we observe. You have little oxygen. It absorbs chemicals. Yes, America's domination. It is strong in development. America is the most powerful continent. You have to be careful about the information you give . Ilona , you can not   to   tell a lot of information. No   tell   everything . Be careful!

They researchers they have also their worries. Many people think that Ilona and Ivana have negativity. We know with whom you're writing, Ilona. We also know that not much security . Is it hazardous . Certain people, no all . . Ivana and Ilona, some people solve this you somewhere other, than the people of the monitor from the computer . Computer as you say, it is not entirely safe. Give Attention ! They someone Solves you from different places in America. But you're in protection from us. We protect you. But it's dangerous. Also, it may fail . Yes, people are different to the understanding and judgment about you. You sit them in the head, those researchers. Ivano, you've got a leg pain that another body in your past, suffered an injury, too. It goes   after life. This is not punishment, but it is the law. We with them to America we have the honor with America long ago. To them you tell. That America is a lot of aliens. The very aliens entity - biological unit. It is very kinds   in America survivors. They survive there in the underground and on the surface. Other races on the basis of a dimension that is hidden from your earth's surface. It's also a lot of dimension to your Earth beneath, underground and on the surface that are not visible to your human senses. You have a different order, system. In every country in every continent are hidden dimensions with other entities. But we do not belong to their other system. Yes, they are also reptilians, as them they say people. They are a different kind. They're annoying, unpleasant. They can sham, hypocrisy and modify change. That is their goal. They are trying to get everywhere. Yes they are everywhere under the surface and on the surface. We do not want, them to have nothing in common with us. They do not like us. It's very kinds of foreign entities from other dimensions, the Galaxy everywhere in the universe. We have another galaxy, another universe. There are so many as the stars and planets from your perspective on your planetary system of the universe. . / What Reptilians ? / - - They are as hybrids between them, but they are too. They knows about himself. They Reptilians to recognize the at a distance. They have secret societies. They are know among themselves. They reptilians live differently. Although they are indistinguishable from humans. Last time was our communication big power. We were last in communication in the vicinity, near. It was also a better connection. Today we are too much far .. We're not in the atmosphere. Next time we will near again, when we will need your samples of everything. We investigate samples of all, what for us signal to determine. Next up will be a signal Ivana will know ... Now we have another task.

Ilona, we can not tell everything. It's a universal law. Ebe end. Next time we communicate.

Aleluja oo !   Hello Bret! Now I'm sending serious information from communications with Ebe - Olie . August 10, 2016. The report also   what the" (New kids)   " I am afraid to publish. But Ebe agrees to publish it for people and researchers who trust us. Or who is interested communication was very fast.. it gave me four hours of work. I translated each word to English , nothing had changed. it is the exact original. I hope it will be understandable.   The whole message is interesting to the end. that's why you'd better read it all somewhere in a quiet place.

Message from humanoid
Carbon Based Ebe - OLie. Telepathic - spiritism communication . Médium Ivana Podhrázská   and   ILona Podhrázská from Czech republic .   "   Message :

18 September 2016

EBE Message: 18. September 2016 – Ebe Communication : Contact - communication 18. september 2016 with Humanoid Ebe - OLie , from planet ELieLji , from 12 dimension . Ivana Podhrazska is spiritism medium and Sister Ilona Podhrazska all writes . from Czech republic Contact was starting 1993 . Message 18. september 2016 : ALeLuja oo ! Welcome ! It is time to time. Time shifted to a certain phase space. You will see things unknown , which eventually understand that you have already experienced. We have the established vibration Regulations allowing us decisions on our navigations , which we have within reach for your government . It is your deity no understand .Ivana a strong personality. Ivana is a goal that is given a certain frequency. You know that when Jesus raised peace, everything changed in the unified moment. People began to doubt his soul. Everything was different . Soul for you he no gave Jesus , but the whole infinity all . Jesus is from another planet. And he is alive. It is 90 light years from your dimension. We have now plan to examine the interior of your forces. The strength of human relate to the whole part central stage tissues a crossroads in the body that forms a nano particle core elements, which forming a higher level of energy development. APROXIMACIE - Just the APPROXIMATION - is simply an explanation of what Olie now told. This is to be formed and developing system other tissues in the human state . Form is the blood of the people, biology, cell particles

and this is the whole and everything is changing every
moment. But the poisons in your environment give
for you result in the conversion and transformation. In
antiquity there was peace in this system . There were poisons,
but chem-trail is future. That was only the beginning. You
live in the future. And the past was the past and worldly
individuals divided and classified according to how one is
preserved. People are structures of the doubt. People
for you see what the human eye is not given. Ivana exudes
a fluid energy balance. The balance consists of JING JANG.
Ivana has JANG. You are the physical state of water, air, fire,
earth. But form only a part of it. A element is a part.
JAKAMA MAKA GAIA. Goal Earth it is the start for no
understandable . Goal is no quality . Manipulation of all
kinds . Your government hiding very important information,
which that is must not get on the surface. They signal and
formed a pact with the government of any country in the
world. But America and China is a big chaos on this branch .
They plan horrific damages commit crimes . We're with
human beings. We are not against the people. They
manipulate human genes along with extraterrestrial entities.
For us it is not an alien, but for us it's a different kind. Chaos.
Some people are poor, some dirt. They poor are
more people. People dirty are atrocity. and it are the
materialists. We see what you think. We this also
now investigating. More people now will be in its central
system, how do you call a brain. A lot of people think that
introducing a different order and people think they are
worried about the future. And the more people know

awakening. Goal of our generation is developing into an all values with human elements. It will be seriously jeopardized human species, when people leave to conquer other realms than we are. Chaos humans occurs when a second large-destroying. And this is already happening with you. Ivana is a strong source of income thoughts . Karma is lost. Each individual is somehow under someone and something that would have led the development of the soul. Never must not no one anyone to influence. It is against fate and against everything that man needs to perform according to their soul. We are now above the astral world. There, people can not see, but can perceive what is happening beneath the surface. Astral world level which is hidden against cleansing in which you suffer. It's a different sphere magnetically connected to the core part of the Cosmos. Broadcasting occasionally systemic signal that sometimes someone feel like vibration, noise, squeaks and sounds of the Earth together. Country form arcs, such as the merger between Astral and your existence. CHAKANA FAJAKA PABAKA . I tis crossroads KABAJA JAKABA ECECHAF. EBE - for you complex expression into for your central we insert. We have lots of plans to create your world to be another. 2 The sun already are. Both manipulation Your government is trying to outshine the Sun a part of your part Nibiru. These two have just intermingled together. It saw the people and it should not this happen properly . And so the government plans to make people not see, the main part of the this happening. It was a great disaster, when the Sun 2 were close together. Some people previously seen as similar to the

planet Earth. It was similar to your country. People panicked, panic and started a war with and objects variously moved across the sky. They wanted their profits. People looked incredulously at it with distaste. People are afraid of themselves. It was a great earthquake mainly in the cosmic flow. The quake was not of the Interior of your country. The earthquake happened in the air. Up there is a lot of things that threaten human civilization. And it will be with you again, it happens. Yes, life in which you Ivano and was seen. Anunnaki arrive there as well, as it was, but in a different location. Anunnaki persecute people. They are tall and changing. Yes, like a civilized dimension ZON. Ivana is also connected to ZON. And we, I do not want to give Ivan Olie into the hands of ZON. The fellow who came to vámv Friday. it's dangerous. He danger. Beauty is not everything . Do not look at anyone with a direct gaze, if you do not know. Ivana is a medium and someone people see auras. He will know the man joker. He Ilona weaken your energy system. You had to feel it. ILona do not do bullshit that can form a game that could get out of hand. You do not know people. Ivana where you lived, was peace. It was beautifully and then a disaster, but it was on one dimension further. Ivana, your soul floated Cosmos Your hierarchy Researcher Bill Forte have question: what it is JALAKA and when extraterrestrials will reveal his identity? / - Identity has already occurred.
We're beings of the Cosmos revealed . JALAKA - is development all blocked bodies, which are the cause of the death of individuals. The body, when the block and ends and the soul remains. Love is a element of the soul. Love

reduces stress and many fun that human body cells and there are flooding the constantly maneuvers. Such as molecules in the speed energy wave .

Researcher Bret Colin Sheppard have questions : Something about the secret world guild ,association ? World association is a guild and world dominion generation. The guild has a ways to different kinds of underground passages and whole tangled , intertwined cities and forts, which is the name MATEBA . / Infinity corridors. And there it is happening session between different governments. Governments do seance. In Arizona, Area 51, Nevada, together with the various entities forming galaxies session. They have a lot of papers on evidence that it transmit for him to him highest. To advance. This is called astral. The government ceded must be done one step, which is very strict because goes is a the earth. Which they want to destroy. Another question, what it is for children with black eyes ? / - Young children have short stature doubtful conscience. They are constituent part of a hybrid. Their genes are manipulated by other state of aggregation of the particles in a development that created the universe. It is a cosmic database - the Law of everything. Children have a material that can not be extradited to the human creature. If the material is ever released, it would be bad. They have eyes hybrids based on experiments from cosmic membership. But they are few. It is only small membership development. We know, yes. / ALenka have question, What will the refugees and new diseases? / - The Third World war is . The refugees are the enemy of humanity. They are handling the reptilians. Disease, which is a large scale is always caused by biological weapons deployed all governments. all country . Because the man himself is a biological entity and anti biology and cell killing everything in the human body. Form enlarged widespread epidemics of all kinds. Yes, OLie ending. OLie goes to the next stage. ALeLuja oo !

22 October 2016

EBE – Humanoid EBE –Olie communication spiritism method 22.October 2016 - EBE - Humanoid OLie from 12 dimension communication method spiritism Message : Ivana Podhrázská
is telepathy contactee , and sister ILona Podhrázská it all writes . Ivana and ILona are in Czech republic . Start contactee with Ebe year 1993 …… English and down is original Czech writing . October 22, 2016 - ALeLuja oo I welcome you today to this your planet Earth. We have you and we are seending you in the corridor of record at some support special message. And this again is a gift to the earth, like England. England is the message deities. We have a record of your history. We are waiting just for reading the time that was given to us. We have many genes in common in like with the Earth England . England's top bodies. Which are in a sense other beings. There are many different structures in England. From your land like Africa, America, Asia. Everything together is MEGLOKA. I give for consciousness . Chaos is happening now with the mutations, which she found herself, and was given to the ancient ties. The atmosphere particle source molecules are based on a scale so inaccessible that it makes , him would marvel prehistoric civilizations. Our source says that England and Russia They are part of the elements conducted on the basis of influencing . Will happen much chaos in Russia and England. America takes care of it. China yes . The atmosphere is greater driver of any kind. Target , goal we have the task of

scanning features of molecules that are in phase disasters. Disaster will , when people start to sleep more. Even if the world wakes up, say, in small quantities, but she wakes up. I give JEMDA is an ancient word opportunity , possibility . Chaos Russia, America. We know a lot of atmosphere, but you know very well that the underground city now forms part of the energy centers that make up the Corridors. They corridors are in the development of molecules and aluminum. Lead up into the atmosphere. Everything is connected . Underground we are investigating and we have samples of the various crap limit. In the underground are forms poison. Like air over you. And people are asleep. spiritual awakening was time, but it was time constraints of the body. The body is under very unreliable. It is linked with the development of various human body. Development did not come here on earth just so. Soul is handling everything and everyone. What do you see, you have it. You must be able to wake up from the bodies. Body ... . And you must improve soul. Its center go to the headquarters of physical bodies. America makes handling this. Forming cells on the basis of common exclusive deformation ecstasy. And . A body can not resist this development. Human DNA is deactivated. In this handling center. Give the investigation. We think even if we do not have a brain. We have, as you say pineal gland. Epiphysis cerebri it is latin name . Our She looks like a flat triangle. We now forces many to decipher the psyche of mankind. Yes, the pineal gland, you say you have it gray and matted, but it gives you data, such as information we give of our bodies to your fields, crop circles which form the shapes of the , divine message. And we examine your pineal gland, and you people have stunted it is creased . We are giving away the message for humanity and you people destroy it. You people do not know the message or you can not understand it. You were not finished works. Energy field in a certain phase space gives many take up . We give in the energy center in your to motherland centers. Your central thread, It is blood . It is giving to our laboratory. 2. We are conducting research your thread on the phase of soft aluminum defect . Aluminum defect affect limitation the structure of your thread in the state body which that your lungs breathe. . It's a defect that affects blood flow. Disaster in the thread that is in your blood results in a very poor reception positives. People should think about your brain. Brain development is in a small part. Since the pineal gland, as I told them. If she were out, so people would not robots. Many common universe .. Take the Sun and the Moon's orbit, and your mother earth. So anyway pulsating heart, body, soul in people. But people have a different structure. We have the two humanoids main in our Ship , who make up a system of molecules and two side humanoids who are just whole watching , monitoring . We Olie watching and is one that comes to the surface and that's the main Humanoid. We can not say everything about the epiphysis cerebri of brain pineal gland. Only it is

clear that her people have gray and hardened, disheveled, destroyed . We just researched the surface of the pineal gland and we took from one kind of earthly man just a sample that was happened to in Dubai. And the man It is respective , honor to the sample . He is weak . We had take took part of the sample. We have his man sample of the energy center . Rid him of tissue . Man is weak . State freely kernel , which has the circulation flow waves into the phase of the blood. Ebe Olie knows that we make investigation for people and kind of underpeople . Mammoths were seconded and it was the message of the soul, as the law says. But mammoths were on the level the Druids. Druids blended with people and not people . .. People were developing hybrids. We no finished higher level, since there was no time. We had a phase in our system to return. And in the future will be the same. History repeats. But the Sun is manipulation. You've got your solar system full of ships that have aftermath of iodine airy realms that are like be simulate . MEGA JAMA say to America, that's a threat to our system. Operations at Sun are bearer GAMA JAMA. Moon is different. He was artificial. Your Moon fell into energy - galactic space. Your Moon manipulation Epileptic. There were operations, as in the Sun. Therefore,

your Sun will also be replaced Nibiru. The second Sun. When it's your will be smaller and weaker, it will be next to Nibiru and brighter. And people will stare at the sky and they will tell what is going on? Second sun is sometimes visible. Already It is / . .... / Eric Mitchell asks why he was contacted, and what they wanted to do? / - DNA contact his mother, so we can contact him, but not we , but our helpers to contact him. / Shaun Coates, whether he is visited by aliens? / - We do not know this person. William Mawers has question how long it will , when contact with whole the world? / - Has long since been. You do not need to have any Lasers monitor sound plates, such like as Alaska antenna. The plates are of many frequencies. These are antennas and have a lot in common with CERN. which is a conductor of these antennas. They are common on the same basis. They together communication.

/Steve Murillo has question what is the free energy of the drive, why is the secret? ... / - Great energy drive is not secret, it is caused just by the energy in the CERN . /

Nancy De Tertre has question is whether you would be able to give a spoken record, to give a sample of their native language and if you have an alphabet ? / - Formulas and secrets and messages, have long had this personality with a big heart long ago decipher in crop cirles there it is. There is a lot of verbal numerical period after period, which show physics, geometry, biology, and especially the Universe. This is on the same
basis .Our speech is not like yours. You have assumption , hypothesis . Simply people minded and do not know. We have a speech on the development of old- Hebrew. Yes, based on Roman - Czech grace. Sanskrit yes.. , MARAHATA CHARARA JAKAMA GAIA. This is influenced air . The monkeys were born
3 to develop 123 OBKALA . That means the "old - Roman font, as it occurred they kind of hybrid development and history continues, but the monkey was human evolution - a monkey. It was still developed from the historical phase . Now there are many other declarations on development that

people will historical people of mutants . But it occurs year 2035 . Some, as we look they are similar like us humanoids and are messengers from the future. Those that will here Earth ., they will comment. And you're sure people subspecies . . But many fruiting. Love, and be in love. Ebe OLie end. We love you, as you also feel us . ALeLuja oo! We also Thank you Ebe, OLie! Thank you so much everyone! Ivana and ILona .

14 January 2017

Aleluja oo! Welcome! You have a new year and with it more worries. We in our planet do not have such a stage. At us time is not. At your planet records of age. While age is just a physical matter. It is the visible universe. It's different than before. The universe is changing because of gravitational energy that comes from the proton. It is a big shift in the antimatter in the Universe. Heavenly bodies changed its trajectory and your country. Nibiru is beside the sun is still long. Nibiru is only hidden, but will slowly uncovers. She and when reveal his one part. The sun fading, but still there will be a few years for your source. Energy still has. Activates its power through charging from ships from other dimensions. It is under the supervision of your Sun and Moon. The moon not hologram, as someone thinks. The Moon is a sort of satellite and the ramp. Your Earth creation using disintegration of the Moon. Now the moon is part of careers. Moon is Educational center for space travels . Moon serves as a base for flights. Hologram are some people on earth. The transmitted through Lazer. He knows the government long ago. You have to pay attention to the people with whom you're talking. . You are talking about us, we know people are talking about you in the world. You must to reach the limits of prudence . EBE is wind . We have the wind in our Ship . The Wind causes make waves signaling in apparatus , which records the evolution of molecules in the atmosphere. We monitor condition in Space . We are now above the atmosphere. We have secured protection against the cold. We like warmth for our ship. It's easier. Your whole family is under pressure from the fear that is unnecessary. We feel it. You must be able to rejoice.Are you a point for handling some people. 10 people manipulate you, but it will protect you, we try. Effort Ivana, you have a physical problem skin leg. It will be well. The pain subsides. Make a wish, it is medicament . We see inside, where the problem does not lie, but they are factors outside. Already we have in our Ship pressure, more chaos . We do not have no more our Ship rays. Chaos climate . / Ebe, it is good that about us was talking about on American radio? / You need to watch out for each other. Not a problem it does not matter. Already happened . We know about it and will all much more. This is just the beginning. But you have to be safety. That's the rule. Our energy is now under pressure absorbed by the Cosmos. Ebe is glad to you have a map of love. We have source love connected around the soul. Part of its development. The surroundings is on record in the program. Capturing discharge aura. The film captures the soul laser beam. The soul is infinite as the universe. Space soul. Strong force hot stripe . Ivana career Goal of our forces is a goal infinity. Our Ship is moving away from the wind. Chaos wind. Maneuvering it with a plate inside a vacuum. It is the density in our Ship . They will be great calamities on earth. Much water will flow to spring. People are talking about us in most states. They do not believe and believe. We are watching it . This source record. We have little worry communication convey different messages. We must be caution . Your reality on Earth takes place much faster.

Humanity perceive time differently as time go by without. Another sphere period come. Everything is different. I already I told . It's a shift. Manipulation of people in a large scale. Ilona, today is the slow communication. Are situations which fluctuation. Yes, at us is shape , completed shape. Final done . The shape of the structure, which monitoring the lives of all entire . The shape of the cells. Ivana and ILona is our strength nice. Nice your doggies, nice to your pets. We do not have animals . We have only a certain type of insect. We have mutated insects . On your Earth is also presence new species of mutant

Ivana with her dog Vanga

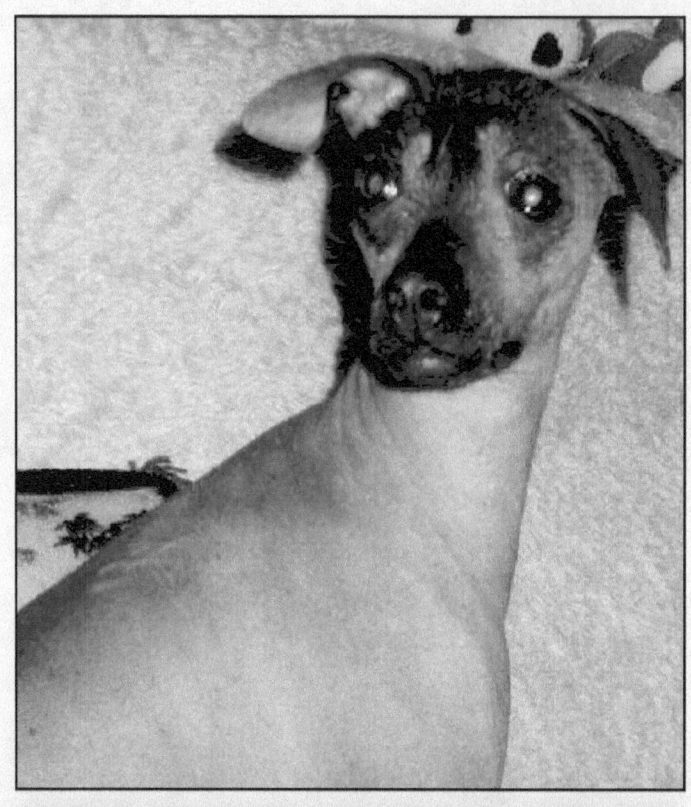

Ilona and Ivana's Doggies (top) Sophie (Bottom) Dancing Queen

Ilona and Dancing Queen

animals and insects. New species and more species acts artificially. A re like Bio-robots a new species of insect. Meanwhile yet occurrence still a low. Other waiting in vacuo to give birth, nativity to their shift. Waiting. Small occurrence in the south of the country. In the northern parts awaiting much more in test tubes. Involved in this is It is caused by your government and with beings from outer space. I can not tell what it is beings. Everything is they monitor. Our signal of our Ship they can monitor, register. This certain kind of beings. Any our manipulation, they may monitor. It's a secret. I know they will many. / Ebe, they destroy us ? / - They do not aim to kill, but another different and much worse things ... I can not now communicate. Maybe i will the next time I'll tell you. EBE must now end. We need to move away much up . was told everything what was given. Good for You be merry! Aleluja oo !

From Ebe - OLie:

Do not like me Olie is, that everyone has an opinion by books. Books also does not know all exactly. Yes in Rooswell we were now 23. september 1993, there because we wanted to be. Previously At that time, our members crashed there. Only one man who gave the message even if no allowed. Neither the aircraft is not registeredthis and you are in the newspapers all sorts of crap, drivel. They wrote journalists everywhere that someone saw us or another.

That is true, but journalists still much to invent

nonsense, Not exactly. I Olie was elsewhere. But our machine, object nobody saw because he put in order and our members ELielji unfortunately stayed there. Nothing more i do not want to do write about it . Only just that plagued him negative creatures from America in a secret laboratory.

Conclusion

We are beings of time. These transmissions are as real as any communication brought to us through mediums who do not know the science or technology involved within the transmissions by this EBE named Olie. These sisters are human beings with no scientific credentials, but with extraterrestrial circumstances that have passionately transcribed these original messages from another world, so that our world may see them, and that we may work together to understand them. These transmissions are reminders of the human spirit showing us that, not only are we not alone in the universe, but that there are humanoids like us sharing the same concerns about the future of all worlds.

Ilona and Ivana would be classically considered Adepts in an ancient society. When the adept is spiritually aligned with any force of our natural world, fire, water, earth, and wind, the Adept can accomplish anything and becomes part of the balance within that force. At that point anything is possible in our custodial bodies. We are just simply part of the Earth for which we have the ability to respond. We are able to use energy for creating new pieces of the whole, as well as starting over after we build our sandcastles of the mind. The manifestation of the sandcastle is less important than the thought, which can never be destroyed by will or any external force. The thought comes from the light and space is there to catch it. We are all potential adepts and are here to Illuminate and educate the future adepts of any world. All humans are potentially adept, and it is with great gratitude that I look to the Czech Girls for insight and friendship between all worlds.

- Bret C. Sheppard -

Appendix

Life in Telč in the Czech Rep.

Maria column, Zachariáše z Hradce Square, Telč, Moravia, Czech_Republic

Ilona and Ivana with their father Joseph

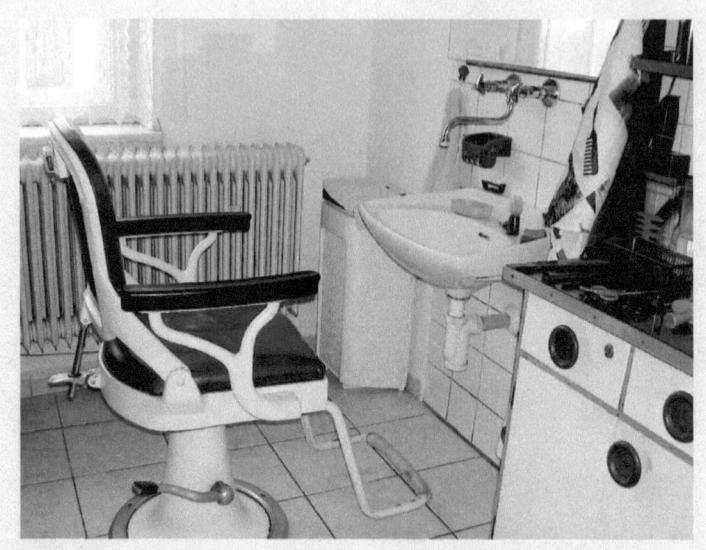

Ilona and Ivana's Beauty Shop

MLUVÍME S MIMOZEMŠŤANY!

Už čtrnáct let komunikují dvě sestry z Telče s ufonem. Ilona je médium, Ivana zprávy zapisuje.

Ilona and Ivana with a new reptilian friend

"Little Grey Friendship"

Happy for you is what I seek.
The colors blind for when we speak.
My friend in peace honorable and true.
The original peace comes from you.
There in the beginning and to the end,
Friends for life together again.

I see your heart in shades of gray.
My alien friend here to stay.
Your ship crashed down,
Like lightning smashed,
Magnetic fields surely zapped.
I'm sorry that they did that to you,
For they knew not what they do.

Were not all like that you will see.
The little gray men in you and in me.
Friends for life we'll always be,
A little Grey friendship for the world to see.

- Bret C. Sheppard -

www.ingramcontent.com/pod-product-compliance
Lightning Source LLC
Chambersburg PA
CBHW030802180526
45163CB00003B/1137